CAMBRIDGE LIBRARY COLLECTION

Books of enduring scholarly value

Life Sciences

Until the nineteenth century, the various subjects now known as the life sciences were regarded either as arcane studies which had little impact on ordinary daily life, or as a genteel hobby for the leisured classes. The increasing academic rigour and systematisation brought to the study of botany, zoology and other disciplines, and their adoption in university curricula, are reflected in the books reissued in this series.

Memoir of the Rev. John Stevens Henslow

John Stevens Henslow (1796–1861), professor of botany at Cambridge University and Anglican clergyman, is best remembered for his role as a mentor to Charles Darwin. First published in 1862, this biography by Henslow's colleague and brother-in-law, Leonard Jenyns, pays tribute to a man he describes as one of the most remarkable of his time. Through vivid accounts of times spent with Henslow both in the university and on travels around Britain, he paints a portrait of a modest and conscientious man, whose pursuits were intended solely for the benefit of others. Recounting Henslow's scientific work and religious endeavours, Jenyns also explores his pioneering contribution to botany and geology, his assistance to the farmers and the poor of his parish, and the role of his faith in his work. Compiled with help from Darwin and other colleagues, Jenyns' memoir provides a unique insight into an important figure in scientific history.

Cambridge University Press has long been a pioneer in the reissuing of out-of-print titles from its own backlist, producing digital reprints of books that are still sought after by scholars and students but could not be reprinted economically using traditional technology. The Cambridge Library Collection extends this activity to a wider range of books which are still of importance to researchers and professionals, either for the source material they contain, or as landmarks in the history of their academic discipline.

Drawing from the world-renowned collections in the Cambridge University Library, and guided by the advice of experts in each subject area, Cambridge University Press is using state-of-the-art scanning machines in its own Printing House to capture the content of each book selected for inclusion. The files are processed to give a consistently clear, crisp image, and the books finished to the high quality standard for which the Press is recognised around the world. The latest print-on-demand technology ensures that the books will remain available indefinitely, and that orders for single or multiple copies can quickly be supplied.

The Cambridge Library Collection will bring back to life books of enduring scholarly value (including out-of-copyright works originally issued by other publishers) across a wide range of disciplines in the humanities and social sciences and in science and technology.

Memoir of the Rev. John Stevens Henslow

Late Rector of Hitcham, and Professor of Botany in the University of Cambridge

LEONARD JENYNS

CAMBRIDGE
UNIVERSITY PRESS

CAMBRIDGE UNIVERSITY PRESS

Cambridge, New York, Melbourne, Madrid, Cape Town,
Singapore, São Paolo, Delhi, Tokyo, Mexico City

Published in the United States of America by Cambridge University Press, New York

www.cambridge.org
Information on this title: www.cambridge.org/9781108035200

© in this compilation Cambridge University Press 2011

This edition first published 1862
This digitally printed version 2011

ISBN 978-1-108-03520-0 Paperback

MEMOIR

OF THE

REV. JOHN STEVENS HENSLOW.

BY THE SAME AUTHOR.

A MANUAL of BRITISH VERTEBRATE ANIMALS: or, Descriptions of all the Animals belonging to the classes *Mammalia*, *Aves*, *Reptilia*, *Amphibia*, and *Pisces*, which have been hitherto observed in the British Islands. 8vo. 13*s*.

THE NATURAL HISTORY of SELBORNE. By the late Rev. GILBERT WHITE, M.A. A New Edition, with Notes and Illustrations. Fcap. 8vo. 7*s*. 6*d*.

OBSERVATIONS in NATURAL HISTORY : with an Introduction on Habits of Observing, as connected with the Study of that Science. Also a Calendar of Periodic Phenomena in Natural History ; with Remarks on the Importance of such Registers. Post 8vo. 10*s*. 6*d*.

OBSERVATIONS in METEOROLOGY ; relating to Temperature, the Winds, Atmospheric Pressure, the Aqueous Phenomena of the Atmosphere, Weather-Changes, &c., being chiefly the Results of a Meteorological Journal kept for Nineteen Years at Swaffham Bulbeck, in Cambridgeshire, and serving as a Guide to the Climate of that part of England. Post 8vo. 10*s*. 6*d*.

MEMOIR

OF THE

REV. JOHN STEVENS HENSLOW,

M.A., F.L.S., F.G.S., F.C.P.S.,

LATE RECTOR OF HITCHAM, AND PROFESSOR OF BOTANY
IN THE UNIVERSITY OF CAMBRIDGE.

BY THE

REV. LEONARD JENYNS,

M.A., F.L.S., F.G.S., F.C.P.S.

" A good man leaveth an inheritance to his children's children."

Prov. xiii. 22.

LONDON:

JOHN VAN VOORST, PATERNOSTER ROW

M.DCCC.LXII.

PREFACE.

In the following Memoir of Professor Henslow, after giving such particulars of his early life as I was able to obtain, I have thrown together subjects of the same kind as much as possible, without regard to a strictly chronological arrangement. It is thought that the work will thus be rendered of more service to those who, from their own particular pursuits or professions, are likely to interest themselves in some of those subjects more than others.

I have derived great assistance, throughout the whole, from Dr. Hooker, and the other members of Professor Henslow's family : indeed, without their friendly co-operation, the work could not have been undertaken. The Professor himself left behind him no journals or other papers that were available for the purpose. The materials had therefore to be gathered entirely from the recollections of his surviving relatives and friends, aided by his own publications, and a few other public documents.

I have not thought it desirable to introduce letters, which would very much have swelled the work, and

in some measure have altered its character ; but I have
given instead extracts from Professor Henslow's published writings, when serving to set forth his opinions
on particular subjects.

The portrait prefixed to the Memoir is a photograph
taken by Mr. Jeffrey, from a bust executed by
Woolner, after Professor Henslow's death.

Besides the members of the Professor's own family,
I am indebted to several other parties for valuable communications to the work, or for help rendered to a
greater or less extent.   Among these I am called upon
especially to mention the names of Mr. Darwin, Rev.
M. J. Berkeley, Professors Sedgwick and Babington,
and Rev. J. Power, Librarian to the University.   To all,
including Professor Henslow's relatives first alluded
to, I beg to return publicly my sincere thanks.

                                    L. JENYNS.
Bath, March 12*th*, 1862.

CONTENTS.

ERRATA.

P. 151, l. 8 from the bottom, *dele* " alone."

P. 154, l. 8 from the bottom, *dele* " which was never published."

P. 165, l. 2 from the bottom, *for* " apprectated," *read* " appreciated."

P. 257, l. 4 from the bottom, *for* " with," *read* " from."

P. 261, l. 3 from the top, *for* " 53.0," *read* " 5.30."

MEMOIR

OF

PROFESSOR HENSLOW.

CHAPTER I.

EARLY LIFE AND EDUCATION.

The death of a man who devoted his whole life to scientific pursuits, and who found his chief happiness in imparting to others the knowledge he had himself acquired, is a public as well as a private loss. Such a man was Professor Henslow, of whom it is proposed to speak in the following pages, and of whom some record of the kind here attempted may have an interest with at least those who are not wholly unacquainted with his name and character.

Few persons in the same sphere of life have exercised so great an influence for good on those about them, or with whom they were associated in the same pursuits. There have been thousands of clergy equally attentive to the duties of their ministerial calling, there have been many men equally eminent in the departments of science. But in Professor Henslow, the clergyman, the

B

philosopher, and the naturalist, were all combined; and though there may have been also not a few clergymen, who like him were fond of science and natural history, yet such individuals for the most part have simply occupied themselves with these subjects in retirement, and for their own private amusement, without any reference to the immediate interests of others.   The striking feature in Professor Henslow was, that every advantage he possessed he gave to the public.   No man ever lived less to himself.   It was this that made him one of the most remarkable men of his day.   Whatever acquirements he made in the various branches of human knowledge, and the degree to which he was master of many of them was very considerable, whatever he took in hand was done with a view to the benefit of all within his reach.   There was no light hid under a bushel; there was no talent laid up in a napkin. This naturally led to his gaining a wide reputation, accompanied by universal esteem.   It led also to the rare success which attended his endeavours to raise the social, as well as the moral and religious character of the neighbourhood in which he lived.   In all the successive periods of his life the same character stands forth.   Whether viewed as a Professor at the University, or as occupying the more humble position of a parish priest, we find one aim and object always before him.   We see the same thirst for science, the same untiring zeal to win others over to a love of the pursuits he so keenly relished himself, the same delight taken in training the young to appreciate and cultivate all truth, and in getting them to take an equal interest in the works and in the Word of God, as forming

different parts only of the great volume put into our
hands to read, and alike calculated to afford lessons
for our growth in virtue and happiness. Gifted himself with intellectual endowments of no common order,
and with an admirable tact for instructing others,—
possessing qualities of the heart which drew towards
him all to whom he had once made himself known,—
blest with a well-regulated mind, totally free from all
selfishness, all vanity and deceit, all love of applause,
all caring for reward, if he could only make himself
useful in the world,—he was the teacher to whom
every one was glad to listen,—the friend whose acquaintance all sought out and valued,—the clergyman
whom his parishioners, though they may not at first
have thoroughly understood him, from the rude and
uneducated condition in which he found them, came in
time to look up to and respect the more the longer he
had resided among them, and in the end to love and
almost idolize, up to the sad hour in which, after a
very short notice, they found themselves bereaved of
one to whose guidance they had freely committed
themselves in all temporal as well as spiritual matters,
and whose like they could scarcely hope to look upon
again.

Other features of Professor Henslow's character
will be brought out in the following pages, in which it
is not intended to give a complete life, materials for
which it would be difficult to collect, but simply a
memoir serving to show what manner of man he was,
and dwelling chiefly on those parts of his career
as a naturalist and a clergyman, which it would be
most useful as well as interesting to record.

JOHN STEVENS HENSLOW was born at Rochester, in Kent, on the 6th of February, 1796. His grandfather, Sir John Henslow, was chief surveyor of the navy, and resided for many years in London in Somerset House. He is said to have " united to a scientific knowledge of naval architecture, great ingenuity and skill in designing;" and no doubt some portion of this ingenuity and skill descended to the grandson, in whom they were equally conspicuous. Sir John Henslow left the Navy Office in 1806, and died at Sittingbourne, in Kent, in 1815.

John Prentis Henslow, the father of the subject of this memoir, practised in early life as a solicitor at Enfield, having succeeded to an excellent business as young as he could take it, in consequence of the death of his Principal by a fall from his horse. He was, however, soon induced to give up his profession and enter into partnership with his uncle, W. Prentis, Esq., largely engaged as a wine-merchant in Rochester. Afterwards, on the occasion of his marrying Frances, the daughter of Thomas Stevens, Esq., a wealthy brewer in the same city, he joined his father-in-law in the brewery business also; Mr. Stevens retiring to his country seat at Gads-Hill House, which he had built some years previously, now the property of Charles Dickens, Esq., so well known in the literary world. In 1808, the partnership between Mr. Henslow and his father-in-law was dissolved, and a few years after Mr. Henslow resumed the practice of the law. In 1828 he retired from active life, removing to St. Alban's, where he died in 1854 at the age of 83.

Mr. Henslow was the father of eleven children, of

whom three died in their infancy. John Stevens was the eldest of all. Three other sons died some years back, though not till after attaining middle life. Four daughters still survive. Young Henslow, of whom we have now exclusively to speak, was sent in the first instance, as a day scholar, at the age of between seven and eight, to a small private school at Rochester, kept by Mr. and Mrs. Dillon, French emigrants; afterwards to the Free School at Rochester, then kept by Mr. Hawkins, head master, and Mr. Benjamin Hawkins, second master. At these schools he received the rudiments of his education. In March, 1805, he was removed to a third school at Camberwell, kept by the Rev. W. Jephson, where he remained, first in the school itself, and afterwards as a private boarder apart from the school, till near the time of his entering the University.

It is always interesting to trace back, where it can be done, to early boyhood, those who have distinguished themselves in any department of literature or science, or who have occupied any prominent place in society. There will often be found little occurrences and facts of no moment in themselves, but worth noticing in connection with the after development of faculties, of which they seem to have been the precursors. A few such incidents are remembered by the surviving relatives of the child, who in this instance grew up afterwards to be the learned Professor. A few words may first be said of his personal appearance. He is described as having been a beautiful boy, with brown curling hair, a fine straight nose, brilliant complexion, soft eyes, and a smile that reached everybody's heart.

A taste for the pursuits to which he gradually oecame more and more attached as life advanced, soon began to appear. He showed his ingenuity, as well as his fondness for natural objects, by making the model of a caterpillar. He would sometimes amuse his brothers and sisters with what he called " gunpowder devils," or astonish them by burning phosphorus, in which we see the first attempt to know a little of chemistry, as well as the desire to open his knowledge to others. His ingenuity was further displayed when very young by producing to his parents a piece of lace made on a biscuit, with bobbins of his own contriving, in imitation of his mother's work on a lace pillow. The passion for collecting was first exemplified by his bringing home the different natural curiosities he met with in his walks. In one instance, while yet a child in a frock, he dragged all the way home from a field a considerable distance off a large fungus,* which when exhibited to the family was jokingly said to be almost as big as himself. This fungus being dried was hung up in the hall of his father's house, and often pointed out to strangers as an indication of the future botanist. And it was characteristic of the turn of mind which so remarkably distinguished him in after life,—a turn for general collecting,—that as a little boy he should have had his attention drawn towards, with the desire to become possessed of, an object belonging to a division of the vegetable kingdom, which few botanists, even at any

---

* This fungus may have been the giant Puffball (*Lycoperdon giganteum*), which is said to attain "often many feet in circumference," or perhaps some large species of *Polyporus*.

period of their lives, care to study, or trouble themselves to collect the forms of which it is made up. We do not wonder, therefore, at the expressions used by some of the friends of the family, in reference to the way in which he thus exercised the rising powers of his body and mind. It was said, "I never saw such an active little fellow ;"—"Nothing escapes that boy's eyes." He was often taken to the house of one old lady, who collected all sorts of curiosities, and who was delighted with his intelligence and looks. He would seem, indeed, to have been generally admired. In childish parties he was the favourite partner whom all courted, and he is said to have danced elegantly.

He may have inherited some part of his taste for natural history from his mother, an accomplished woman, who, though she never studied it as a science, was, like the old lady above alluded to, a great admirer as well as a collector of natural and artificial curiosities. His father, too, was extremely fond of birds and other animals, as well as of his garden. At one period of his life his father had an extensive aviary, comprising a great variety of species, some of which are not often seen in cages in private houses. His library also contained a good many books on Natural History. This was quite enough to create an interest for such things in the child, while the taste thus excited was, as might be supposed, duly encouraged by the parents.

It was not, however, till after going to the school at Camberwell that his taste for natural history began to develope itself in a more scientific manner. Mr. George Samuel, the drawing-master at that school, was himself an entomologist. It may be supposed, there-

fore, that the latter soon took an interest in the young
pupil, whose taste was the same as his own, warmly
encouraging as well as assisting him in his pursuits.
Young Henslow was often seen running about an
orchard adjoining the school with his green gauze
butterfly-net; and occasionally the drawing-master
and he went out to longer distances insect-hunting
together. From the drawing-master he learnt the art
of setting his insects and preserving them properly.
His eldest sister well remembers, on his coming home
for the holidays, the astonishment with which she and
her parents stood gazing at the beautiful collection of
insects, as it appeared to them, in the drawer at the
bottom of his school-box,—where they were pinned
down on pieces of cork curiously stuck together,—
and listening to all he had to say about them.

It is believed that it was through the drawing-
master at Camberwell, and about this time, that he
made the acquaintance of two other naturalists who
still further assisted him. These were no other than
the celebrated Dr. Leach of the British Museum, and
Mr. James Francis Stephens, the celebrated entomologist.

Dr. Leach was thoroughly master of every branch
of zoology, and Mr. Stephens had the best collection
of British insects in the country. Naturalists of this
stamp and standing were not likely to let the young
collector's zeal evaporate. They at once fixed him
down to the pursuits, which had been hitherto taken
up as a mere boyish amusement, but which were
henceforth to be made regular studies. The woods of
Kent were now well searched for insects, while the

Medway was explored for shells. He was always active and always busy. He had, indeed, before this, shown a desire to become acquainted with the inhabitants of the water as well as the inhabitants of the air. His father was in the habit of making yearly picnic excursions with his family up the Medway, and the boy was delighted with the opportunity thus afforded him of fishing for all he could get, while the family were enjoying themselves in a very different way. He has been heard to say that these were the happiest days of his life, though he generally got well scolded by his parents for spoiling his clothes.* The fruits of his industry, however, now acquired more value from his superior knowledge of the science. Among the acquisitions made to his collections were many interesting and little-known species,—crustacea and shells from the Medway and the adjoining salt-marshes,—lepidoptera, of which few specimens previously existed in cabinets. Some of these being shown and given to his patrons

* This, perhaps, will not be wondered at, when it is mentioned that among the shells which he obtained from the Medway, either at this time or at a later period, were *Mya arenaria* and *Mactra compressa* of Montagu. A fine series of each of these species is in my own cabinet, for which I am indebted to him. Of the former, Forbes and Hanley in their "British Mollusca," observe that, "although locally abundant, and tolerably diffused throughout our coast, it is less frequently met with, especially in fine condition, than might be imagined ; being dug out of a gravelly, sandy, or clayey bottom near low-water mark, usually in or near estuaries!" Of the latter, that "it principally inhabits sludge or muddy places, buried to the depth of five or six inches ;" adding, "it is from the comparative inaccessibility of such spots, that the species, although most abundant, is not frequently taken alive, and that cabinets are usually only furnished with dead valves, washed on shore after rough weather."

Dr. Leach and Mr. Stephens, the habitats were recorded by those gentlemen on the authority of the
captor, * and the specimens found a place in their
respective drawers now in the British Museum. A few
years later, when at Cambridge, he forwarded to Dr.
Leach from that neighbourhood, a second species of
Paludina that had not before been recorded as British,†
besides several entirely new small fresh-water bivalves ;
and in 1817, he sent him from the north coast of
Devon, where he also collected one summer, a new
crab, which from its peculiarities Dr. Leach considered
as *sui generis*, and described and figured it as such. ‡

It was while still at the school at Camberwell, and
ardently engaged in his natural-history pursuits, that
an incident occurred, which, had the desire to which it
led been carried out, might have thrown him into a
totally different sphere of life from that to which he
was destined. He was not so engrossed with natural
history as to neglect the studies of the place, and for
his proficiency in the latter he obtained a prize. The
prize proved to be Levaillant's "Travels in Africa."
Levaillant was a naturalist who had collected and described many of the animals found in that large and
little-explored country, and young Henslow, from the

---

* *Corophium longicorne.* Of this small crustacean, Samouelle,
who wrote from Dr. Leach's manuscripts, says in his "Entomologist's Useful Compendium," p. 105, " very common at the
mouth of the river Medway, where it was first observed by
J. Henslow, Esq." Mr. Stephens, in his notice of *Setina irro-
rella,* a moth described as " very irregular in its appearance,
and excessively local," says, " taken copiously near Rochester
in 1816 and the following year, by Rev. Professor Henslow."
—*Illustrations of Brit. Ent. Haust.,* vol. ii., p. 99.

† See *Leach's Brit. Mollusc.,* pp. 208 and 291.

‡ *Polybius Henslowii.* See *Malac. Brit.,* t. ix. B.

time of reading this book, was filled with the idea of treading in the steps of the illustrious traveller. He felt a strong desire to become a traveller himself, and in the same country. The thought of the wide field that would be thus opened to him, where he might give full scope to his passion as a collector and a naturalist, explore regions but little known, and enrich science with new species in every department of zoology, engrossed his mind. As might be supposed, the idea was warmly taken up and encouraged by Dr. Leach, who rejoiced in every project that was likely to promote the interests of natural history, as it was also by some other friends; but, alas for him, not by his own family, who were utterly opposed to it. For how long he indulged the hope of being an African traveller is not exactly known; but it is believed to have been either in 1810 or 1811 that he got the above prize, and it was not till many years afterwards that he was at length persuaded to abandon the scheme upon which he had set his heart. During this interval, it is said that his mother had many anxious moments, from the pertinacious way in which he clung to the idea of going out. He himself, too, often came home depressed and out of spirits. And even long after he had given in to the wishes of his relations, he still continued to think much upon the subject, read with the greatest interest many other African travels that were published from time to time, and the volumes procured to gratify this taste, continued to occupy a place on the shelves of his library to the day of his death.

There can be no question that had he been allowed to follow this calling, he was admirably fitted for

it. Strong in health and powers of endurance, of
a robust habit of body, possessed of great courage, and
not easily turned by dangers where there was anything
in view, with a zeal and love for science that few could
exceed, and a mind and eye always well directed for
seizing every advantage, and securing everything that
was new, it would be not easy to say what was wanting to have made him an accomplished and successful
traveller. Yet those who believe in an over-ruling Providence will not regret that he was held back in his
designs. They will lean to the persuasion that higher
ends were to be carried out by him, and a greater
good to be secured to others, if not to science as well,
by his settling down in the quiet neighbourhood where
he spent so great a part of his days, and where his
mind and faculties were never for a moment less active
than they would have been in the wilds of Africa itself.
He was, indeed, one of those few persons, who, by a
rare combination of qualities, are fitted for almost any
sphere of life in which their lot is likely to be cast.

It was in October, 1814, that Henslow was entered
at St. John's College, Cambridge. Though devoted
to Natural History, he did not allow himself to be
drawn away from mathematical studies, which at that
day were the prevailing studies of the place (especially
at the college to which he had joined himself), to the
exclusion of many subjects which now form part of the
academical course. His powers of reasoning and clear
faculties well suited him for studies which necessitate
much application of the mind and close thought. Nor
do mathematics, which are needed for the successful
cultivation of some of the sciences in which he took

interest, appear to have been at all distasteful to him. The result was, that on taking his degree of B.A. in January, 1818, he obtained the fair place of sixteenth wrangler, the senior wrangler of that year being Lefevre, afterwards Clerk of the Parliaments. At the same time, he did not neglect the teaching that was to be had by attending the public lectures of the professors. Whilst an undergraduate he acquired a tolerably correct knowledge of the two cognate sciences of chemistry and mineralogy. Dr. E. Daniel Clarke, the celebrated traveller, was at that time professor of mineralogy, and Cumming, professor of chemistry, and under these eminent men he learned enough of the two subjects they respectively taught to enable him a few years afterwards to become a candidate for one of those chairs himself. To geology he does not appear to have paid any particular attention till after taking his degree. It was that same year, 1818, that Sedgwick, now a veteran in the science, was appointed to the Woodwardian Professorship, by the resignation of his predecessor, the Rev. J. Hailstone. Mr. Henslow was introduced to him through Lefevre, the senior wrangler of his own year, and who was a Trinity man like Professor Sedgwick, and during the Easter vacation of the following year (1819), he accompanied the Professor in a tour to the Isle of Wight, where he took his first lessons in practical geology.

Mr. Henslow might have read more or less on the subject of geology before going to the Isle of Wight, but reading books and working with the hammer in the field are very different things; and it is alike creditable to tutor and pupil that the latter should have been so ready a learner, and so soon have mastered the details of stratification and the distinguishing characters of rocks, as to have been equal to exploring for himself, and with his own unassisted eye and hand, another district very shortly afterwards, and even to have gathered fruits worthy of being laid before the public. This district was the Isle of Man, which he visited during the long vacation of 1819, in company with some Cambridge students, whom he took with him as pupils. What leisure time he could spare from attending to their tuition, he devoted to the geology of the island, still, however, collecting specimens in zoology (some of which were forwarded to his friend and patron Dr. Leach), as well as a few of its rarer plants. Its geology, indeed, was not entirely unknown. Dr. Berger had previously published an account of the island in one of the earlier volumes of the Geological Transactions, but Mr. Henslow was able to correct a few mistakes which that author had made, as well as to point out the localities of several formations unnoticed by him. These additional remarks he embodied in a paper, illustrated by a map and sections, entitled " Supplementary Observations to Dr. Berger's Account of the Isle of Man," which he contributed to the Transactions of the same Society in 1821. He had been elected a Fellow of the Geological Society in 1819, and Fellow of the Linnean Society in 1818. And this was his first essay in authorship at the early age of twenty-five.

He returned from his tour to the Isle of Man in high spirits, and full of the adventures he had met, as also of the kind and friendly attentions he had received

from Dr. Murray, at that time Bishop of Sodor and
Man, who took great interest in his researches, sometimes accompanying him in his walks, and who presented him, on leaving, with a walking-stick, made of
the fossil oak found in the peat-bogs of the island, or,
as they are there called, curraghs. He was also much
pleased from another circumstance that had occurred
of more importance to science. He was fortunate in
having been in the Isle of Man the very year in which
the first perfect skeleton of the so-called " Irish Elk "
was found in that island. With that zeal for science,
and for the interests of his own University, which so
distinguished him through life, he did all he could to
secure the prize for the Woodwardian Museum. In
this he did not succeed, as it went eventually to the
Museum at Edinburgh. But he was so bent upon
having it, if possible, and so unwilling to give the
chance up, that after his return to Cambridge, he
actually went back to the Isle of Man in severe winter
weather to renew the attempt, fruitless as it turned
out. I am indebted to Professor Sedgwick for these
particulars. The disappointment, indeed, was of no
lasting consequence to the University. An equally
fine specimen was procured by purchase from Ireland
a few years afterwards, and this latter forms now one
of the chief ornaments and most valuable of the
treasures in that grand collection of geological and
palæontological specimens, the Woodwardian Museum.

After quitting the Isle of Man, Mr. Henslow continued to give close attention to geology, and, encouraged by what he had done in the above field, soon
entered upon another which called for higher powers

of philosophical research. This was the Isle of Anglesea, to which he bent his steps in the summer of 1821,
accompanied, as before, by pupils. He was here not
only without a fellow-worker in the geological line, but
it was a mountainous region, the geology of which was
of a very complicated character, and which had never
been satisfactorily worked out by any previous observer.
Perhaps those only who have laboured in a similar field
can estimate all the pains that are required for such an
investigation,—the task of disentangling the truth
from the many sources of error which arise to perplex
the observer,—the watch that must be kept for faults
and other causes of dislocation of the strata,—the
difficulty of determining the relative age of many of
the older beds, in which there are few or no fossils,
and where the geologist must be guided chiefly by
their mineral characters,—in short, the patience with
which the whole ground must be gone over again and
again to insure correctness. Mr. Henslow brought all
his knowledge of chemistry, mineralogy, and mathematics, and his skill in drawing, in which he excelled,
to bear upon the inquiry. He mapped the island, and
after having thoroughly secured its physical features,
as well as mastered its geological details, the results
were incorporated in an elaborate and most valuable
paper, read to the Cambridge Philosophical Society in
November, 1821, and published in the first volume of
that Society's Transactions. This paper, which was
accompanied by an admirable collection of the rocks of
Anglesea, now in the Woodwardian Museum, secured
for its author a high reputation as a geologist. Its
merits were widely appreciated, the more so consider-

ing his age at that time, only twenty-six, and the limited experience he had gained by previous researches. It may be mentioned, as one proof of the value set upon it by English geologists, that the volume in which it appeared was speedily out of print, mainly, it is believed, from the great demand for the memoir in question.

Having mentioned the Cambridge Philosophical Society, it is necessary to go back a few years in order to speak of the origin of that institution, which Mr. Henslow was one chief party in getting established. It was during his tour in the Isle of Wight with Professor Sedgwick, in the spring of 1819, that the two fellow travellers talked the matter over together, and started the project. The first idea was to establish a Corresponding Society, for the purpose of introducing subjects of natural history to the Cambridge students. The matter was not forgotten, and soon after their return to the University, they commenced writing to their respective friends for their encouragement and support. Sedgwick wrote to Professor Jameson, of Edinburgh, and to some of his friends in Cornwall, he himself visiting Cornwall in the long vacation of that year. Henslow wrote to his friends in London on the same subject. Thus the spring and summer were spent in taking the necessary steps for making their scheme known. When the University met again for the October term, the joint founders of the new institution (that was to be) called upon Dr. E. D. Clarke, who immediately entered warmly into the project, and it was agreed to call a public meeting of the members of the University in the lecture-room under the library, to take into consideration the formation of a society for promoting the study of natural history. The following is the notice issued, under the date of October 30, 1819.

" The resident members of the University, who have taken their first degree, are hereby invited to assemble at the lecture-room under the Public Library, at 12 o'clock, on Tuesday, November 2, for the purpose of instituting a Society, as a point of concourse, for scientific communications."

This notice had thirty-three signatures attached, comprising the names of many of the heads of houses, professors, tutors, and others. The meeting took place as appointed, and the following resolutions were carried unanimously :—(1.) " That Dr. Haviland be called to the chair. (2.) That a Society be instituted, as a point of concourse, for scientific communications. (3.) That a committee be appointed to report to the members of the University, desirous of belonging to the said Society, such regulations as shall appear to them to be proper for the proposed institution."

The above committee met November 8, to consider of the rules ; and at a subsequent meeting, held November 15, their report, together with a code of regulations, was read and approved of. Sedgwick made the motion for its adoption, which being acceded to, those present voted themselves a Society, and the institution, under the name of the " Cambridge Philosophical Society," dates its establishment from that day.

At the same meeting, the officers and council of the Society were appointed. Professor Farish was elected the first president, Professors Lee and Sedgwick the first secretaries. It was arranged that the ordinary meetings of the Society should "be held on a Monday, once in every fortnight, during full term." The first meeting took place in the Museum of the Botanic Garden, on Monday, December 13, 1819, when the rule declaring the objects for which the Society was instituted was altered, to stand as follows:—"That this Society be instituted for the purpose of promoting scientific inquiries, and of facilitating the communication of facts connected with the advancement of philosophy and natural history."

Henslow was, at that time, only a Bachelor of Arts, or he would have had a prominent office assigned to him. He, however, became one of the secretaries in May, 1821, and continued to hold that office till November, 1839, when he ceased to reside in the University, up to which period he continued one of the Society's most active members.

This Society, which owes so much to him, still thrives, and is now (1861) in the forty-second year of its existence. It numbers, at the present time, upwards of 500 members, has a good museum, especially of British animals, a fair library, and has published nine volumes of Transactions, in which are some papers of first merit and value, and equal to any that are to be found in the publications of similar institutions elsewhere.

It was, I think, in the year 1820 that I had the happiness of making my first acquaintance with Henslow. He, at that time, occupied lodgings in All-Saints Passage, since pulled down to make room for the new court at Trinity, erected by the present master, Dr.

Whewell. Among his most intimate acquaintance, and with whom he seemed most to associate, were Calvert, then tutor of Jesus,—Dawes, of Downing, the present Dean of Hereford,—Kirby, of Clare,—and Ramsay, of Jesus, who was next Henslow on the Tripos. All these are now dead, with the exception of the second. Henslow was busily engaged at that period in arranging his cabinets of British insects and shells, both which collections he presented to the Cambridge Philosophical Society, the year after it was instituted, and for which he received the thanks of the Society on the 13th November, 1820.

He also devoted some part of his time to mineralogy and chemistry. He had several drawers of minerals by him, and a quantity of chemical apparatus, which, along with his collections in natural history, had a small room to themselves above his sitting-room. When I attended Professor Cumming's lectures that same year, Henslow assisted him in the lecture-room, by which he must have gained much practical knowledge of that branch of science. Zoology, however, seemed then to be his favourite pursuit. Botany had scarcely been taken up, otherwise than by collecting a few plants in the Isle of Man, as before mentioned, though, within a twelvemonth from that time, we commenced together the formation of a collection of British plants. He would often make excursions in the fens or down the river, where he was the first to notice many of the local species of insects and shells that are to be found in that district. There is one little creek running into the Cam, at a place called Backsbite, about two miles below the town of Cambridge, where he was fortunate in discovering, upon floating weeds, two specimens of an insect which, from its extreme rarity, has been met with only in one or two other instances in this country. This species, the *Macroplea equiseti* of entomologists, is figured in Curtis's "British Entomology," from the identical specimens taken by him on the above occasion; and the volume in which it occurs is dedicated to the captor, by its talented author, in terms which afford high testimony to the opinion the latter entertained (and no one could be a better judge than he in this matter) of the zeal uniformly manifested by Mr. Henslow towards the advancement of British Entomology. His zeal was scarcely less shown in the department of British Conchology. The same creek in which the *Macroplea* was found was the spot in which he first detected some of the small freshwater bivalve shells before alluded to, and which, till his time, had been confounded together, so far as they had been noticed at all, and thought to be only the young of the common *Cyclas cornea*, found in almost every piece of water throughout the country. These latter captures were shown to Dr. Leach, by whom it had been intended to describe and publish them in the work in which he was then engaged, on the Mollusca of Great Britain, but which work never appeared till many years after the death of the author. One of these shells, the most remarkable of them all, from the little winged appendages attached to the upper part of the valves, had the name of *Henslowiana* given it by the Doctor, which name it has ever since continued to bear.

The discovery of these small bivalves is the more

worth noticing, from its being due to the circumstance
of a most useful net which Henslow invented about
the year 1815 for straining the minutest shells from
the mud of ditches, in which many of them are found
deeply embedded.   Conchologists had previously made
use of iron spoons or ladles with holes in them, which
did the required work very imperfectly.   The net here
spoken of is constructed of the finest wire gauze bound
round an iron frame, one or two inches deep inside,
with a second netting of coarser wire outside to defend
it from injury : a wooden handle may be affixed of any
length, or a common walking-stick made to screw
into it.   It serves the purpose for which it is intended
admirably; and it is only one of the many instances
in which through life Henslow displayed his ingenuity
in contriving whatever he wanted from time to time to
assist him in carrying out his designs.   He was always
ready with some mechanical contrivance to meet difficulties and to further his operations.

I am unwilling to speak of myself; but it was
entirely by the help of this net, and the co-operation
of its inventor, that many years after, when I had
discovered one or two other species of these little bivalves, I was enabled to publish a "Monograph on
the British Species of Cyclas and Pisidium," a contribution to British conchology which Henslow would
have himself previously brought out, but for the circumstance of his attention being drawn to more important objects of research.   I feel, indeed, here
called upon to state how much I owe to his friendship.
I had, long previously to making his acquaintance,
devoted myself to the study of natural history, but he

was the first with whom I could ever associate in the same pursuits, and but for him what little I have done for the science would have been still less, and might have been nothing. He was always ready to encourage, to advise, and to assist. He was the means of introducing me to many other naturalists in London and elsewhere, as well as getting me admitted to some of the Natural-history Societies. In Cambridge, when first I knew him, though there were men eminent in other branches of science, there was hardly a single resident member of the University besides himself who cared for zoology. We naturally, therefore, associated much together, and often made expeditions for collecting about Cambridge, and in the neighbourhood of my father's house at Bottisham Hall, where he was a frequent visitor, and which led ultimately to his connection with the family as hereafter mentioned.

Those who have gone on for many years alone following up such pursuits can best appreciate the advantage and satisfaction of enjoying the companionship of others of the same taste as themselves; the spur thereby given alike to the physical and mental powers, by which things are attempted and projects carried out, which would otherwise never have been undertaken. As regards myself I may truly say that some of the days thus spent in company with Henslow were not only among the most enjoyable of my life, but the most profitable in respect of all that I learnt from him not merely on natural-history subjects, but on others with which he was equally conversant. At one time we meditated bringing out jointly a Fauna Cantabrigiensis, and with this object in view we rambled, on different occasions, to some of the more distant parts of the country, such as Ely, Wisbeach, and Gamlingay, spending a night or two at each place. It was in August, 1824, that we made our first visit to Gamlingay, a locality so rich, from its geological position, in rare species of insects and plants not found in any other part of Cambridgeshire, and so well known to Cambridge men in later times, from the excursions yearly made there by Henslow with his botanical class, after being appointed to the botanical professorship. We returned home laden with spoils,—pleased especially with the discovery of the natter-jack in great plenty on Gamlingay Heath, a species of toad that had only been found before in one or two spots in England, and of which we brought away many living specimens. Gamlingay had been previously searched for plants by Relhan, the author of the Flora Cantabrigiensis, but we were perhaps the first naturalists from Cambridge that had visited the district since his time, and obtained its other productions.

The contemplated Fauna of Cambridge never appeared, partly from Henslow's time being fully occupied with the duties of the two professorships which he successively held in the University, and of which it is now necessary to speak, and partly from the circumstance of myself ceasing to reside in the county before the work could be got ready for publication. Considerable materials towards it, however, were acquired, which may yet perhaps be one day availed of by some one willing to undertake and complete the task which we only commenced.

CHAPTER II.

In the early part of 1822 Dr. Edward Daniel Clarke, Professor of Mineralogy, died. Mr. Henslow was a candidate for the vacant Professorship. He was young: he had only incepted the year previous, and had not yet completed his M.A. degree. But there were few, perhaps, at that time among the resident members of the University better qualified for the office than himself. He had a good knowledge of mineralogy and the allied science of chemistry. He was also generally esteemed. There were certain circumstances, however, connected with his election which had the appearance of opposition to his appointment, and to which it is necessary to make some allusion. This Professorship had been founded by a grace of the Senate in 1808, on purpose for Dr. Clarke, who for two years prior to its establishment had given lectures on mineralogy, illustrated by a large collection of minerals he had made during his travels. On Dr. Clarke's death the University purchased his collection of minerals, and determined that the Professorship should be continued. A

c

dispute, however, arose as to with whom rested the power of election.   The Heads of the Colleges claimed the right of nominating two candidates for the Senate to elect one of them; while the Senate asserted that the election should be by open vote, *more Burgensium*, without any previous nomination by the Heads.   This led to protracted litigation.   There was no real desire that any other than Mr. Henslow should have the appointment.*   He was one of the two candidates nominated by the Heads, and on the day of election he had a larger number of votes recorded for him than their other candidate.   He was consequently declared duly elected, and the Vice-Chancellor admitted him to the Professorship.   But the Senate generally would not accept him.   A great majority of the members being determined to try the question as to their right of electing whomever they chose, tendered their votes in favour of a third individual, who, without any wish for the office, simply allowed himself to be brought forward for the occasion, with the intention of resigning in favour of Mr. Henslow, whenever the right in question should be established.   On demand being made that this third candidate, who had obtained the greatest number of votes, should be admitted to the Professorship instead of Mr. Henslow, the Vice-Chancellor refused.   Application was then made to the Court of King's Bench for a mandamus to admit him, and the case was several times argued, but not decided, when in May, 1823,

* This dispute gave rise to several pamphlets on the occasion, in one of which, by Professor Christian, it is said that Mr. Henslow "had no rival candidate," and that he (the writer) had "never heard the slightest breath of hostility expressed against him."

circumstances occurred which put a stop to the further discussion of the case in a court of law. The question was afterwards referred by the Senate to the arbitration of Sir John Richardson, ex-Justice of the Common Pleas, but his decision was not given till December, 1827, before which time Mr. Henslow had ceased to hold the Professorship.

Immediately on getting this appointment, Mr. (now Professor) Henslow gave his whole mind to the subject of mineralogy. He entered upon its duties under a disadvantage in one respect. His predecessor, Dr. Clarke, had been a very popular lecturer. His lecture-room was sometimes so crowded that it was difficult to get a seat. This arose mainly from the eloquence of his style, and an impressive way he had of fixing the attention of his hearers and getting them to take an interest in the subject before them. But he added to the popularity of his lectures by widening his field, and sometimes introducing topics, in themselves attractive, but somewhat foreign to pure mineralogy.* Professor Henslow's style was not what would be called eloquent, but he had a good voice and a remarkably clear way of expressing himself, and of explaining in well-chosen language anything that his hearers might

* " The plan which the Professor (Dr. Clarke) pursues is in some particulars peculiar to himself. Besides the usual information on the subject, it contains remarks on the natural history of the various materials which have been adopted, both in ancient and modern times in *architecture* and *sculpture*, and professes to elucidate the knowledge possessed by the ancients of mineralogy, as it is displayed in the Sacred Scriptures, or in the writings of the Greeks and Romans."—*Wainewright on the Literary and Scientific Pursuits encouraged and enforced in the University of Cambridge*, 1815, p. 57.

find difficult to understand. This was what he peculiarly excelled in as a lecturer. Indeed, in after years,
when he had become more accustomed to lecturing,
and was in the habit of addressing audiences of various
kinds, he so thoroughly acquired the art of adapting
his language to the capacities of those who sat under
him, and making himself intelligible on almost all
subjects, that he became one of the very best lecturers
of his day. In his lectures on mineralogy he confined
himself to the characters, uses, &c., of the simple
minerals, as distinguished from aggregate masses,
which belong rather to geology. He noticed even but
very slightly certain mineral masses of homogeneous
texture, such as obsidian, basalt, jade, &c., which do
not form true species. Mines and mining, which
formed the subject of one of Dr. Clarke's lectures, he
wholly omitted, these being fully treated of in another
course given at that time by Professor Farish on the
arts and manufactures connected with chemistry.

He was not long in preparing a Syllabus of his
lectures, of such a character as to show plainly how
completely he had mastered the science in a very short
time. He paid great attention especially to the crys-
tallographical part of it, in which he was much assisted
by his previous training in mathematics when an undergraduate. His Syllabus was published in 1823. It
was drawn up with great care, and was very different
from that of his predecessors. It was so devised as
to serve for more general purposes than as a guide to
the lectures. Stating at the commencement the different heads under which the characters and descriptions of the simple minerals may be arranged,

there is then given a systematic list of the genera and
species, to each of the latter being annexed a table of
its chemical analysis, when known, together with its
" specific gravity and other properties, which require
numerical exposition." This renders the Syllabus a
useful manual of reference to all persons studying
mineralogy, independently of the immediate circumstances which led to its publication.

But Professor Henslow's abilities were about to be
exercised in another department of scientific research
and usefulness. In 1825 the Professorship of Botany
became vacant by the death of the Rev. Thomas
Martyn, B.D. This branch of natural history had, on
the whole, more attractions for Professor Henslow than
mineralogy. Indeed, the Professorship of Botany was
the one to which he had been looking for some years,
and for which he had been preparing himself at a time
when he never anticipated that the Chair of Mineralogy would be open to his acceptance first. He
offered himself immediately as a candidate for it. He
was elected in the room of Professor Martyn without
opposition, and he retained this second Professorship
for the remainder of his life. He did not formally
resign the Professorship of Mineralogy till three years
after. This, however, was simply because Sir John
Richardson's determination as to the mode of election
to that Professorship had not yet been given. But he
virtually resigned it, and in 1828, soon after the settlement of the question just alluded to, Mr. (now Dr.)
Whewell was regularly appointed to succeed him in
that office.

It may here be observed that the Professorship of

Botany at Cambridge formerly, and up to the time of Professor Henslow's death, included three distinct offices. First, the University Professorship, without an endowment; secondly, the Regius Professorship, with an endowment; and thirdly, Walker's Lectureship. All three were combined in Professor Martyn, and they would naturally have been combined in his successor also; but owing to the circumstance of the mode of election to the University Professorship being, at the time of Professor Martyn's death, still under consideration by Sir John Richardson, to whom the question had been referred, together with that relating to the Professorship of Mineralogy, no appointment to the first took place. Consequently Professor Henslow was only Regius Professor, this appointment resting with the Crown; and Walker's Lecturer, who is appointed by the Governors of the Botanic Garden. At his death (I am informed by his successor, Professor Babington), the Regius Professorship, which was made for the second Martyn, became extinct, and the endowment was transferred to the University Professorship, which was vacant during all Professor Henslow's time, but is now the only Professorship existing.

The same Governors who appoint to Walker's Lectureship, had formerly the entire management and control of the Botanic Garden, which, as well as the Lectureship, was founded by a Dr. Richard Walker, formerly Vice-Master of Trinity College. For some few years, however, the Garden has been directed by a Syndicate, consisting of the Governors and six elected members of the Senate. There is no necessary connection between the Professorship of Botany and the

Garden Establishment, and it is only "by the Professor also enjoying the office of Walker's Lecturer (without salary) that he becomes entitled to make use of it." These two offices, however, always go together as a matter of course.

As may be supposed, Professor Henslow took a warm interest in the Botanic Garden, though he had no power to effect any changes in it otherwise than as allowed by the Governors. He found it, when he entered upon the Professorship, if not in a neglected state, at least in a condition "utterly unsuited to the demands of modern science." It was in the heart of the town, very confined, and without possibility of enlargement. It did not consequently admit of more than a limited number of species being cultivated in the open ground, while the houses for greenhouse and stove plants were too small and too few in number. The whole garden was in the same state in which it had probably been for a long period back. He often reported its inefficiency to the Governors, but it was many years before he could get anything done, or the necessary funds raised for remodelling it. At length an opportunity occurred in 1831, by which the University was enabled to purchase a more-extensive piece of ground for a new garden, consisting of about thirty acres, in the outskirts of the town, though from unavoidable circumstances many years more elapsed before the actual transfer of the plants took place. In the meanwhile Professor Henslow published, in 1846, an "Address to the Members of the University of Cambridge, on the expediency of improving, and on the funds required for remodelling and supporting, the

Botanic Garden." In this tract he expresses a desire
to see the Garden "raised to a level with some other
establishments of the same kind," observing "that the
larger the number of living species that are cultivated
in a Botanic Garden, the greater will be the facilities
afforded—not merely for systematic improvement, but
for anatomical and other experimental researches essential to the progress of general physiology. It is impossible," he says, "to predict what particular species
may safely be dispensed with in such establishments,
without risking some loss of opportunity which that
very species might have offered to a competent inves-
tigator, at the exact moment he most needed it. The
reason why a modern Botanic Garden requires so much
larger space than formerly, is chiefly owing to the
vastly-increased number of trees and shrubs that have
been introduced within the last half-century. The
demands of modern science require as much attention
to be paid to these, as to those herbaceous species
which alone can form the staple of the collections in
small establishments."

The year previous to the circulation of this "Address," he had, "at the request of the Trustees [more
correctly Governors], taken some pains in assisting
them to secure the services of a Curator, who should
be competent to meet the demands which such an
establishment" as he hoped to see in the University
might require. "At their request also, and that of
the Syndicate" appointed to consider the subject of
the Botanic Garden, he "twice visited the National
Establishment at Kew, and obtained from its talented
and experienced superintendant, Sir W. J. Hooker, as

well as from Dr. Lindley, and some others thoroughly
acquainted with the details of Botanic Gardens, such
information and advice as were calculated to put the
Garden upon at least an equal footing with those of
Edinburgh, Glasgow, or Dublin."

The result of these inquiries was the gradual preparation of the piece of ground above spoken of for
the purposes of a new Garden, and the securing the
services of a skilful and well-qualified Curator, the
late Mr. Murray (who unfortunately died in 1850),
by whom, in conjunction with Professor Henslow, the
arrangement of the Garden was determined upon and
carried out, and " a greatly-enlarged space appropriated
to the cultivation of the plants." Very little alteration,
Professor Babington tells me, has been since made in
the plan then agreed to. Botanical science was the
first thing considered; "but, in order to encourage a
general taste for botanical studies, and to render the
Garden an agreeable acquisition to the University, the
designers consulted ornamental appearance, whenever
it did not interfere with the main object." The first
tree was planted in October, 1846, by Dr. Tatham,
then Vice-Chancellor, just before he went out of office.
Only the trees were planted in that year, and not all
of them; it requiring several years to get the requisite
funds for planting the rest of the trees, as well as for
stocking the beds with hardy shrubs and herbaceous
plants. " The trees form a belt surrounding the whole
of the ground, and are arranged as far as possible on a
scientific plan, so that allied genera and species are
near to each other. Amongst them will be found
nearly all the trees that will stand our climate, and it

is believed that, when grown up, they will form one of the most perfect Arboreta in the kingdom." The herbaceous plants form "a very perfect and valuable collection," which is said to be "much used by the members of the University who study scientific Botany."

In the formation of the new Garden, the chief object has been the cultivation of hardy plants, and in that respect it has fully succeeded. The funds at disposal do not at present allow of much house-room for the tender kinds. The plant-houses hitherto erected were constructed, in 1855, upon a plan suggested by Mr. Stratton, the Curator (successor to Mr. Murray), and have proved very convenient. "They are capable of containing a sufficiently numerous collection to be tolerably illustrative of the chief groups of the plants that inhabit the warmer regions of the earth; and such, to a great extent, is the collection which is now successfully cultivated in them. It includes a considerable number of fine specimens of the older inhabitants of green-houses, now rarely to be seen, and also many very interesting species which, from possessing slight claims to beauty, have fallen out of cultivation." *

To return to Professor Henslow's first appointment to the Chair of Botany. I remember his telling me, soon after he got the Professorship, that though Botany was the subject he had long preferred, he felt so much interest in mineralogy, after having followed it

* First Annual Report of the Botanic Garden Syndicate, dated March 5, 1856, from which some of the previous statements are also taken.

up closely for three years, that he thought he should always continue to give up one day in the week to it if practicable. The mind sometimes, when well possessed with one study or pursuit, can hardly believe for the moment that it will ever be so completely occupied with any other as to lay the first aside. Abandon mineralogy entirely Professor Henslow never did. He kept it up to a certain degree, so as not to forget what he had learnt, and so as to be always able to bring his knowledge into use when needed. But he soon found himself more and more absorbed by the duties of his new Professorship. He found his botanical work continually increasing; his leisure for other things fast diminishing.

And in truth he had plenty to do from the very first. He had to begin with studying the science of Botany more deeply than he had hitherto done. For though he had been long looking to the Botanical Chair as what he might perhaps one day occupy, he tells us himself that, "when appointed to it, he knew very little indeed about botany, his attention having before that been devoted chiefly to other departments of natural history;" adding, however, that he "probably knew as much of the subject as any other resident in Cambridge." * He had next to prepare notes for his lectures. And yet further, he had to look over and arrange a considerable collection of dried plants left to the Botanical Museum by his predecessor in the Chair. These had been lying for many years in a damp closet, neither cared for nor attended to, and

* Address to the Members of the University of Cambridge on the subject of the Botanic Garden, 1846, p. 16.

when examined were found greatly decayed. Whole packets consisted of little more than plain paper,—the entire specimen having been in some instances eaten away by insects, or small fragments only of the original plant remaining in others. At the day in which these plants were dried, no one ever thought of applying the poisonous mixture which is now so successfully employed to preserve herbaria from mould and the attacks of insects. Professor Henslow immediately set to work and applied this mixture to all the specimens that were worth keeping, and rearranged the whole in a methodical and systematic manner.

This collection forms but a very small part of the treasures accumulated by himself during the period of his holding the Professorship, and now finding a place in the Botanical Museum. Many valuable collections of plants from foreign countries were obtained by gift or purchase, or through exchange with other Botanists and other Botanical establishments. One rich and extensive collection was the legacy of the late Dr. Lemann. Professor Henslow himself added to the Museum a nearly perfect collection of the plants of Great Britain, the fruits mostly of his own industry.

And besides the herbarium, with indefatigable zeal he gradually got together, for illustration of his lectures, an extensive and most valuable collection of a miscellaneous character, such as I have the authority of Dr. Hooker for saying exists nowhere else. It is difficult to enumerate all that this collection contains ; dried and beautifully-mounted specimens of parts of plants, where characterized by any peculiarity of structure ; diseases of plants, whether caused by fungi,

insects, parasites, or abnormal growth ; similarly-pre-
pared specimens of their accidental variations; products of plants applied to economic uses; fruits,
seeds, and specimens of woods; type specimens dis-
sected out and illustrated with drawings ; diagrams of
various kinds; notes of the prominent members of all
the natural families of plants ; together with statistical,
physiological, medical, commercial, and historical details, respecting many of the more interesting species.
All these (as everything he had) were so arranged as
to be capable of being turned to at any moment, and
brought out for use and exhibition when wanted.

In entering upon his first course of lectures, Professor Henslow had to draw the attention of his
hearers to a branch of science which had long been
neglected in the University. His predecessor had
held the Professorship for the protracted term of
sixty-three years. He was a very old man when he
died, and from age and infirmities had long previous
to his death ceased to lecture, or even to reside in the
University. There had been no lectures on Botany
given in Cambridge for at least thirty years. At one
time Professor Martyn "deputed an able botanist,
a member of the University, to read lectures in his
stead;" but it is said that "these lectures did not
succeed, having, for some reason or other, been but
slightly or not at all attended." * Afterwards, in
1818, the late Sir James Edward Smith came forward
to supply the deficiency, and applied for leave to

* Considerations respecting Cambridge, more particularly
relating to its Botanical Professorship. By Sir James Edward
Smith. 1818. p. 6.

deliver a course of lectures on Botany in the University. He had the consent of Professor Martyn and the then Vice-Chancellor, but there was such strong opposition from the tutors of the Colleges, on the ground of his being neither a member of the University nor a member of the Church of England, that he was forced to withdraw. Professor Henslow set to work with the same energy he displayed when first appointed Professor of Mineralogy. There were some who still remembered Professor Martyn's lectures, but their recollection of them did not convey a very favourable impression of their character. I heard one individual say that they were the dullest and heaviest things imaginable. Professor Henslow had already had three years' experience in lecturing, and he neglected nothing in his power to make his lectures attractive and popular, without departing from a plan of study that could alone give the students a proper grounding in Botany as a science. One great assistance he derived from his admirable skill in drawing. His illustrations and diagrams just now alluded to, representing all the essential parts of plants characteristic of their structure and affinities, many of them highly coloured, were on such a scale that when stuck up they could be plainly seen from every part of the lecture-room. He used also to have "demonstrations," as he called them, from living specimens. For this purpose he would provide the day before a large number of specimens of some of the more common plants, such as the primrose, and other species easily obtained, and in flower at that season of the year, which the pupils, following their teacher during his explanation of their several parts, pulled to pieces for themselves. These living plants were placed in baskets on a side-table in the lecture-room, with a number of wooden plates and other requisites for dissecting them after a rough fashion, each student providing himself with what he wanted before taking his seat. In addition to these there were rows of small stone bottles containing specimens of all the British plants that could be procured in flower, the whole representing, as far as practicable, the different natural families properly named and arranged.

In this manner Professor Henslow not only gained a hearing for the subject of Botany, but he induced many to take up the science in good earnest, and to study it in further detail by themselves afterwards. Some who never thought of the pursuit before, continued to follow it up for several years while a few are still labouring i n the field in which he first set them to work. He had a very gratifying attendance at his lectures for the first seven years, his class numbering from more than sixty to nearly eighty active students; after this, there was a falling-off, due, probably, in great measure, to the additions made about that time to the examination for a common degree, which left the men less leisure for studying natural history. In 1852 and the years following, when (as will be more fully spoken of afterwards) there was an enforced attendance on Professorial lectures by all taking an ordinary degree, who had not commenced residence earlier than October, 1849, the numbers again increased, some years to what they were at first. Many graduates, however, in addition to the junior members of

the University, were to be found among his hearers.
Even ladies sometimes petitioned to be admitted, or
stealthily admitted themselves.

Professor Henslow published his first Syllabus in
1828. This was followed a year after by a Catalogue
of British Plants. The general plan of his lectures
was, after some introductory matter, first to explain
the structure of vegetables, the cellular and vascular
tissues, cuticle, sap, &c.; then to treat of the conservative and reproductive organs, stem, root, leaves,
flower, and fruit; and after that to pass to the process of
germination in the three leading divisions of Dicotyle-
donous, Monocotyledonous, and Acotyledonous Plants.
The history and detailed description of each organ in
succession was next entered upon, with its several
modifications in different cases, especial notice being
taken of any peculiar forms or appearances which it
assumes in certain families. This arrangement of the
subject, though for the most part kept to, was occasionally varied a little, and much improved in details
from year to year, as new matter came to hand, and
as he himself gained more experience in ascertaining
what was requisite to make his lectures profitable.

These lectures were delivered regularly each year
during the Easter term. Towards the latter part of
the course, when the season was more advanced, the
Professor usually made herborizing excursions with
his class, which tended greatly to keep up the interest
of his pupils in the subject, as well as to give them
that practical knowledge of plants in the field which
needs to be added to lecture-room teaching in order
to make men good botanists. These excursions became

in time very popular. From Professor Henslow's extensive acquaintance with all branches of Natural History, and the delight he took in imparting information to all who sought it, he was joined, not merely by botanists, but by entomologists and others, who, through his friendly encouragement, had been led to take up natural history in one or other of its departments. A few came, who, though not naturalists themselves, enjoyed the day for its own sake, the healthy exercise it afforded, the varied knowledge they had the opportunity of picking up, and the agreeable society of those in whose company it was spent. The excursions were either to the Fens, or some other district in the neighbourhood of Cambridge; or to Gamlingay, which last place being more distant, an attractive spot for collectors as already mentioned, and little visited, drew larger numbers, and was the excursion most looked forward to in the season. To convey the party, a stage-coach with its four horses was usually hired for the occasion, and well filled. The appearance of such a vehicle, with the influx of so many strangers, in an obscure village, naturally excited at first some astonishment among the natives, and puzzled them much as to the object of their coming. Nor did the sight of the nets and boxes serve to enlighten them, till they had become familiarized with the uses to which these implements were put. They then learned to take part themselves in the fun of the day as they esteemed it; and, after a time, made a sort of holiday of it, being always ready to greet the arrival of the wise men as soon as they had notice of the intended visit, and in hopes also of making a little money by the sale of any rare specimens they had been fortunate enough to pick up.

Professor Henslow did not neglect his own advancement in the science of Botany, any more than he did that of his pupils. Having thoroughly acquainted himself with the systematic part of it, he lost no time in studying vegetable physiology, and those laws of morphology, which often afford the only key to the real relationships of plants. The extent to which he made himself conversant with the former of these subjects, may be gathered from an able Review he wrote of the Vegetable Physiology of De Candolle,* in which he fully discusses some of the leading points with which all our knowledge of the true functions of the organs of plants is inseparably bound up. His remarks also in the same review on "species," and "hybrids," and the "individual," serve to show the philosophical manner in which he could approach the consideration of those vexed questions. He had previously to this, in 1831, published in the "Transactions of the Cambridge Philosophical Society," a valuable paper, illustrated with plates, in part coloured, from his own accurate drawings, "On the Examination of a Hybrid Digitalis," by which it will be seen how careful he was to inquire for himself on this subject, when the opportunity offered, and how competent to conduct the closest investigations into the character and circumstances of those minute organs with which hybridity is connected.

Dr. Hooker observes, that "this paper belongs to a

* De Candolle's work was published in 1832. Professor Henslow's article will be found in the 22nd No. of the *Foreign Quarterly Review.*

class of botanical papers of which there are unfortunately too few, but which are perhaps the most important of any in a biological point of view, in respect *i. e.* of the physiological significance of structural modifications, or changes, effected by crossing. It is very much an experimental inquiry, and has been quoted up to this day as a model of scientific research. The illustrations, too, are very skilful, the analyses (microscopic) admirable."

There was a second paper published by him in the same Transactions in 1833, " On a Monstrosity of the Common Mignionette." This, too, " is an admirable specimen of careful and accurate observation, quick appreciation of the very obscure relations of the parts of the flower, and excellent judgment. The point he investigated—the real nature of certain organs—was long a disputed one amongst the first botanists of Europe, and his solution was complete, accepted by all, and remains unassailed."

These two papers, both of which Dr. Hooker has pronounced to be " of the highest merit as works of philosophical research," served to establish Professor Henslow's reputation amongst continental naturalists.

In 1836 he put forth further results of his studies, by publishing an introductory work, entitled " The Principles of Descriptive and Physiological Botany." This little work, which formed originally one of the volumes of Lardner's Cabinet Cyclopædia, reached two editions, and a third was in preparation at the time of his death, with extensive modifications. It was long considered the best manual of structural and physiological botany in the English language: the morphological part in particular was excellent. Indeed, for many years there was scarce any other work of the kind in this country that dwelt on these subjects with any ordinary weight; or if there were, Henslow's excelled them all for clearness, conciseness, and simplicity.

Independently of the able manner in which he has treated all the simpler parts of the science, and the accurate wood-cuts which are scattered throughout this volume, he has handled in a masterly way certain of the more-complicated problems connected with the structure of plants. As an instance of this may be mentioned his remarks on Phyllotaxis, or the spiral arrangement of foliaceous appendages, in his successful explanation of which he showed how he could bring his knowledge of mathematics to bear on any question as wanted. This subject seems to have been peculiarly congenial to him. About the time when he first took it up he read a lecture upon it at one of the evening meetings of the Cambridge Philosophical Society, illustrated by a large drawing of a fir cone, having all the relative numbers affixed to the different scales in the order of their arrangement, similar to the reduced wood-cut in the above work. From the calculations and measurements which it involved, the subject afforded much interest to the class of hearers he had in such a place, and who were not, perhaps, previously prepared to find any connecting points between botany and geometry.

In the same volume, under the head of " Colours of Flowers," he displayed his ingenuity by the construction of a Chromatometer for estimating the different degrees of mixture between the three primary colours of red, blue, and yellow; or, between their binary, &c. compounds, which go to make up any particular colour which it is wished to describe. The reader is referred to the volume itself for more full details on this subject, which was also explained to the same Society on another occasion and gained the attention it deserved.

Many years later Professor Henslow published two other useful little works on Botany. One was "A Dictionary of Botanical Terms," illustrated with numerous wood-cuts, in which there is "a ready reference to any term that may be met with in botanical authors, whether it be still in use, or has become so far obsolete, that even proficients in the science may be at a loss to ascertain its meaning without more trouble than they would be willing to bestow." Great pains have been taken to mark the different senses in which different authors have sometimes employed the same word, as also the cases in which they have attached the same meaning to different words. This Dictionary was originally commenced, but not completed, in "Maund's Botanist," and "Maund's Botanic Garden," portions appearing at a time in the successive numbers of those works, to which Professor Henslow contributed. It was first published separately in 1857, forming a volume of a convenient size for the pocket, and since then, there has been a second edition called for.

The other work above alluded to is of a more local character; a "Flora of Suffolk," undertaken in conjunction with Mr. Edmund Skepper, of Bury St. Edmunds, and published only the year before his death.

CHAPTER III.

On the 16th of December, 1823, the year after he was appointed to the Professorship of Mineralogy, Professor Henslow laid down his bachelor life, and married Harriet, second daughter of the Rev. George Leonard Jenyns, of Bottisham Hall, in the county of Cambridge, and sister of the writer of this memoir. The following year he took orders.

His parents had always been desirous that he should go into the Church, and he is said to have acquiesced in their wishes; yet he does not seem to have fully determined on this step till some years after he was of the proper age. It has been stated by a Unitarian connection of the family, that the reason why he did not take orders earlier, was from religious scruples; he could not reconcile the doctrine of the Trinity as held by the Church of England, with what he considered to be the true teaching of Scripture. If this be the case, which I do not remember to have heard him mention, and should hardly think likely, it is quite certain that his scruples were soon overcome, and that he never entertained a doubt of the truth of this doctrine in after-life. It is more probable that he delayed for a time, in consequence of his mind being much taken up with the various scientific pursuits in which he was engaged; while, so long as he continued a bachelor, there was no immediate occasion to add to his income. He waited, also, till some fair prospect opened for him in Cambridge, without his having merely to work for others, as he would have been liable to be called upon to do, if he had had no regular duty of his own. After his marriage, he felt it necessary to delay no longer, at the same time that he had the offer of the curacy of Little St. Mary's, one of the churches in Cambridge, which, from the time of his ordination, he continued to hold till he got higher preferment. He was ordained deacon by Dr. Sparke, Bishop of Ely, in the parish church of St. George's, Hanover Square, London, on the 11th April, 1824; and priest by the same Bishop, at Ely, on the 7th November of the same year.

These two events, his marriage and his ordination, in connection with his professorship, fixed his residence in Cambridge for the subsequent fifteen years of his life or more. He now gave up his lodgings, and took a house suited to family requirements, which became eventually the familiar resort of all whose pursuits at all assimilated with his own.

Professor Henslow was always happy, always active in the discharge of life's duties, and always enjoying life's innocent pleasures. It would perhaps be wrong, therefore, to set one period of his life above another in these respects. Yet he never afterwards appeared

to so much advantage, or seemed so completely in
his right element, as during the time of his residence
in the University. It afforded exactly that kind of
society to which he was best suited, and gave him
just those employments in which he found most pleasure. He was, indeed, constrained for a time to take
pupils, which drew upon his leisure, but he was free
from the harassing cares of a large parish. And
though it was with admirable success that he sur-
mounted all the difficulties he met with, and by the
most praiseworthy efforts that he accomplished all the
good he did, at Hitcham in after-years, as in due
course we shall have to notice, yet it was through the
power he possessed of accommodating himself to circumstances, and giving his whole mind to any work
that duty required of him, that these results were
brought about. At Cambridge, his circumstances were
none other than what he would have chosen for himself, and never have wished to change, had his Professorship, which was nominally £200, but from which
he only netted £182, been sufficiently endowed to
make him wholly independent of church preferment.
He had attained what I believe to have been his highest ambition, the Chair of Botany, and it opened to
him the field of exertion he best loved, and in which
he could afterwards work to his heart's content. His
engaging manners and friendly disposition, added to
the position he occupied, caused him also to be
looked up to and esteemed by a large circle of friends
and admirers. Alike desirous of instructing and
pleasing, ever setting forth the attractions of his own
favourite science, and opening its treasures to all who

would come and see, he became the centre of a sphere, within which all were drawn whose hearts at all responded to the call he gave them. It is difficult to estimate the degree to which his influence extended in forming the characters, as well as in directing the tastes and pursuits of the undergraduates of that day. Nor was it merely by the popularity of his lectures that he effected so much that was good and valuable in this way. Another circumstance operated perhaps more powerfully than his lectures, in bringing it about. This was a practice which he established, and long carried on, after he had become settled in the town as a married man, of receiving at his own house, one evening in the week, all who took the slightest interest in scientific, and especially natural history studies. Doctors and other graduates, as well as the junior members of the University, might be sometimes seen at these meetings. As in the case of his botanical excursions, it was the varied knowledge and accomplishments of Professor Henslow that brought so many together, though they might not all have a leaning to the same pursuits he had. But every one might learn something from him, and every one seemed to go away pleased. At the same time, it was for the benefit of the younger students that these meetings were mainly in the first instance got up. He would seek out any that were reported to him as more or less attached to natural history, while he made converts of not a few who were thrown accidentally in his way. If any young man, through timidity or reserve, shrunk at first from going to the Professor's house, under the idea that the Professor

D

and the Don went together, the frank and open-hearted welcome he received from him when once prevailed upon to attend his *soirées*, soon inspired confidence, and put him at his ease. No Professor in his address and conversation, ever had less of stiffness or formality than Professor Henslow; there was nothing but the name of a Professor to create alarm or keep back.

It may be confidently stated that never before the period we are now speaking of was natural history so much in favour in the University; nor has it ever since held the place it then occupied. There are some, now in the first rank of living naturalists in this country, Darwin, Berkeley, Lowe, Miller,* Babington,† and others, who were then students at Cambridge, and foremost among the followers of Professor Henslow to profit by his teaching and example. Few were more regular than some of these in their attendance at the evening meetings above spoken of. What rendered these meetings the more attractive to naturalists was a display of specimens of minerals, plants, and insects, or any other objects of interest that had recently been acquired by the Professor. Those also who came were invited to bring specimens that were likely to afford interest to the party assembled. In this way it may be easily supposed, kindred spirits soon made each other's acquaintance, while they afforded each other mutual assistance in their respective pursuits. So numerous, indeed, were the Entomologists in particular at that period in the University, that several persons among the lower classes derived a part of their liveli-

* The present Professor of Mineralogy at Cambridge.
† The present Professor of Botany : he succeeded Henslow.

hood during the summer months from collecting insects for sale, especially in the fens which abound with so many rare and local species.

As it is the main object of this memoir to set forth Professor Henslow's character in a true point of view, and to show the influence for good he exercised on others,—it gives me great pleasure to be able to insert the following recollections of him from Mr. Darwin, the first of the names above mentioned, and thus to have much of what I have stated confirmed by one who knew him so well, and who could so thoroughly appreciate the excellence of his disposition :—

" I went to Cambridge early in the year 1828, and soon became acquainted, through some of my brother entomologists, with Professor Henslow, for all who cared for any branch of natural history were equally encouraged by him. Nothing could be more simple, cordial, and unpretending than the encouragement which he afforded to all young naturalists. I soon became intimate with him, for he had a remarkable power of making the young feel completely at ease with him ; though we were all awe-struck with the amount of his knowledge. Before I saw him, I heard one young man sum up his attainments by simply saying that he knew everything. When I reflect how immediately we felt at perfect ease with a man older and in every way so immensely our superior, I think it was as much owing to the transparent sincerity of his character, as to his kindness of heart ; and, perhaps, even still more to a highly remarkable absence in him of all self-consciousness. One perceived at once that he never thought of his own varied knowledge or clear intellect, but solely on the subject in hand.   Another charm, which must have struck every one, was that his manner to old and distinguished persons and to the youngest student was exactly the same: to all he showed the same winning courtesy.   He would receive with interest the most trifling observation in any branch of natural history; and however absurd a blunder one might make, he pointed it out so clearly and kindly, that one left him no way disheartened, but only determined to be more accurate the next time.   In short, no man could be better formed to win the entire confidence of the young, and to encourage them in their pursuits.

" His Lectures on Botany were universally popular, and as clear as daylight.   So popular were they, that several of the older members of the University attended successive courses.   Once every week he kept open house in the evening, and all who cared for natural history attended these parties, which, by thus favouring intercommunication, did the same good in Cambridge, in a very pleasant manner, as the Scientific Societies do in London.   At these parties many of the most distinguished members of the University occasionally attended; and when only a few were present, I have listened to the great men of those days, conversing on all sorts of subjects, with the most varied and brilliant powers.   This was no small advantage to some of the younger men, as it stimulated their mental activity and ambition.   Two or three times in each session he took excursions with his botanical class; either a long walk to the habitat of some rare plant, or in a barge down the river to the fens, or in coaches

to some more distant place, as to Gamlingay, to see
the wild lily of the valley, and to catch on the heath
the rare natter-jack. These excursions have left a
delightful impression on my mind. He was, on such
occasions, in as good spirits as a boy, and laughed as
heartily as a boy at the misadventures of those who
chased the splendid swallow-tail butterflies across the
broken and treacherous fens. He used to pause every
now and then and lecture on some plant or other
object; and something he could tell us on every
insect, shell, or fossil collected, for he had attended
to every branch of natural history. After our day's
work we used to dine at some inn or house, and most
jovial we then were. I believe all who joined these
excursions will agree with me that they have left an
enduring impression of delight on our minds.

"As time passed on at Cambridge I became very
intimate with Professor Henslow, and his kindness
was unbounded; he continually asked me to his house,
and allowed me to accompany him in his walks. He
talked on all subjects, including his deep sense of
religion, and was entirely open. I owe more than I
can express to this excellent man. His kindness was
steady: when Captain Fitzroy offered to give up part
of his own cabin to any naturalist who would join the
expedition in H.M.S. *Beagle*, Professor Henslow recommended me, as one who knew very little, but who,
he thought, would work. I was strongly attached to
natural history, and this attachment I owed, in large
part, to him. During the five years' voyage, he regularly corresponded with me and guided my efforts;
he received, opened, and took care of all the specimens

sent home in many large boxes; but I firmly believe
that, during these five years, it never once crossed his
mind that he was acting towards me with unusual and
generous kindness.

"During the years when I associated so much with
Professor Henslow, I never once saw his temper even
ruffled. He never took an ill-natured view of any
one's character, though very far from blind to the
foibles of others. It always struck me that his mind
could not be even touched by any paltry feeling of
vanity, envy, or jealousy. With all this equability of
temper and remarkable benevolence, there was no in-
sipidity of character. A man must have been blind
not to have perceived that beneath this placid exterior
there was a vigorous and determined will. When
principle came into play, no power on earth could
have turned him one hair's breadth.

"After the year 1842, when I left London, I saw
Professor Henslow only at long intervals; but to the
last, he continued in all respects the same man. I
think he cared somewhat less about science, and more
for his parishioners. When speaking of his allotments, his parish children, and plans of amusing and
instructing them, he would always kindle up with
interest and enjoyment. I remember one trifling fact
which seemed to me highly characteristic of the man :
in one of the bad years for the potato, I asked him
how his crop had fared; but after a little talk I perceived that, in fact, he knew nothing about his own
potatoes, but seemed to know exactly what sort of
crop there was in the garden of almost every poor man
in his parish.

" In intellect, as far as I could judge, accurate powers of observation, sound sense, and cautious judgment seemed predominant. Nothing seemed to give him so much enjoyment, as drawing conclusions from minute observations. But his admirable memoir on the geology of Anglesea, shows his capacity for extended observations and broad views. Reflecting over his character with gratitude and reverence, his moral attributes rise, as they should do in the highest character, in pre-eminence over his intellect.

" C. DARWIN."

Mr. Berkeley also, the eminent cryptogamic Botanist, has been kind enough to send me some remarks on Professor Henslow's scientific habits and scientific abilities, which I conceive will prove of equal interest with the foregoing. He writes as follows :—

" For the last thirty-two years I have seen very little of Professor Henslow; about twice at Dr. Hooker's, and some three or four times at Cambridge. My more intimate acquaintance with him was from 1821 to 1829.

" I had for some years before I went up to Cambridge been much attached to Botany, and Henslow called upon me, at the request of some relations at Blackheath, who had in former days been acquainted with certain members of his family. At that time he was making Botany a study with a view to the Regius Professorship, rather than from any peculiar predilection for that branch of natural history above other branches. His habits, however, were so methodical that he had no difficulty in taking up half a dozen

subjects at the same time, and he had so many little
plans and schemes for economizing time and commu-
nicating efficiently with others, that he could get
through more work in a given time than perhaps any
of his contemporaries. During my residence at Cambridge, I could scarcely be said to be on very intimate terms with him, but after I took my degree, I
went two or three times to visit him ; once, I recollect,
with my friend Lowe, and profited very much by our
intercourse. I had the pleasure of being present at
one or two of his *soirées,* which he contrived to make
extremely agreeable and interesting ; and he not only
gave young men of a scientific turn opportunities of
beneficial intercourse, but he had especial meetings of
more intimate friends for improvement in French.
Somewhere about this time he was engaged in his
history of a hybrid Digitalis, which is a model of
patient investigation. He had a good microscope,
though of rather early date, and was well skilled in its
use.

" After this, I was for some years much engaged
with clerical duties, and lost sight of him, though we
occasionally corresponded, especially on the diseases of
corn, and he was certainly one of the first, if not the
very first, to see that two forms of fruit might exist in
the same fungus. He at once ascertained that *Uredo
linearis* was a mere juvenile condition of *Puccinia
graminis,* and believed that *Uredo rubigo vera* was
only a form of the same thing, many parallel cases to
which are given by Tulasne and others in more recent
times. He was, in fact, so completely free from prejudice, and so desirous of ascertaining the truth, and

nothing but the truth, that he was the man of all others for such investigations, and I have always regretted that he gave so little time in after-days to original researches. His labours, however, as a lecturer, and above all as a clergyman, were directed with such success, that their effects will not be lost during this or the next generation, and investigations which would have redounded to his credit as a botanist have fallen into other hands, which were utterly unequal for the task which he accomplished. He attained perfection in his own especial vocation, and we have, therefore, no right to regret that he followed out the course which was indicated by Providence, rather than one which we might have chosen for him ourselves."

It has been already observed, that during the chief part of the period of his residence in Cambridge, Professor Henslow was obliged to take pupils, from the straitness of his income as a married man. This employment took up much of his time, and it is a marvel that under such disadvantages he could find leisure to do so much in other ways. He himself observes that "five or six hours a day devoted to cramming men for their degrees is so far apt to weary the mind as to indispose it for laborious study, especially if we happen not to be gifted with talents or energy sufficient to overcome such an obstacle." * He here, in his usual modest way, ignores the fact of his having possessed any such talents or energy as he alludes to, though assuredly without such endowments

* Address to the Members of the University of Cambridge, p. 17.

he never could have attained the high position in
science he afterwards occupied.

These pupils were generally undergraduates at some
of the colleges; but occasionally he received boarders
into his house, either preparing for the University, or
for the sake of his teaching independent of a university education.

It was in company with two pupils of the latter
kind, that he and his family went to Weymouth
during the long vacation of 1832, and took a house
there, which he continued to occupy some months.
Greatly did he seem to enjoy the relaxation afforded
by such a trip, and the new field it opened for scientific
researches. Portland, of course, was often visited, and
its geology explored; while he collected freely the
marine animals which are to be found so plentifully in
Weymouth Bay, and on the adjoining shores. He
seemed at that time, and just for the season, to have
busied himself almost entirely with zoology rather
than botany. He kept several fishermen in pay to
bring him the refuse of their nets, from which he got
together a large and valuable collection of crustacea,
mollusca, star-fish, echinidæ, corallines, and other
marine productions; many of which were placed afterwards in the museum of the Cambridge Philosophical
Society, while others were given to myself and different
friends. The disinterested way in which he would
collect for others, or give them the best of his specimens, without retaining any for himself beyond duplicates—a feature this which particularly characterized
him throughout his whole scientific career—was never

more displayed than on this occasion. It was at the
time when I was collecting materials for my manual of
British vertebrate animals, published a few years afterwards; and knowing that I was anxious to obtain
examples of all the British fishes that could be easily
procured, I cannot say to what cost or trouble he did
not go in order to make as perfect a collection of
the species of fish found on that coast as he possibly
could. He was constantly keeping up the attention of
the fishermen to bring him especially any of the rarer
kinds that might fall into their hands, while he was
ever on the look-out himself in the market, or when
the boats came in. Yet he knew little or nothing of
ichthyology himself, he had never attended to it
before, and he took no further interest in the subject
afterwards. It was solely to assist me in the work in
which I was engaged; and the many valuable specimens he brought home, all carefully preserved in
spirits, may be seen to this day in the museum of the
Cambridge Philosophical Society, where, after having
served my purpose for examination and description,
they were finally deposited.

On the 12th of November, after his return from
Weymouth, at one of the meetings of the Cambridge
Philosophical Society, he gave an account, illustrated
by drawings and diagrams, of the various observations
he had made in geology and natural history during his
residence there, having previously delivered a lecture
on the same subject to the people of Weymouth themselves.

It was in the early part of the same summer of
1832, in which he was at Weymouth, that Professor

Henslow had the offer from the Lord Chancellor, at that time Lord Brougham, of the Vicarage of Cholsey-cum-Moulsford in Berkshire. This living, which is set at the value of £340 per annum, and which was, consequently, a considerable addition to his income, he at once accepted. He did not, however, leave Cambridge entirely, simply passing the long vacations among his parishioners while he continued to hold that living,—which he resigned on getting the living of Hitcham,—and residing for the remainder of the year in the University.

Before speaking of the time in which, a few years afterwards, he quitted the University altogether as a residence, it is necessary to make some allusion to a subject, quite foreign to natural history, which for a certain period occupied much of his thoughts, and led him into a course of action, which, if some of his friends disapproved of, they could not all the same but admire the conscientious sense of duty, as he judged it, under which he acted. I allude to the decided line he took in politics in 1835.

Professor Henslow was originally a Conservative, and a supporter of Lord Palmerston, who, for many years, was one of the members for the University. When Lord Palmerston changed his politics on the going out of the Wellington administration, after the accession of William IV., and joined the new reform ministry, Professor Henslow changed with him. Of course, the latter laid himself open to some attacks by this tergiversation, which he bore most good-humouredly, and without the least flinching. "As he had always been a firm supporter of Lord Palmerston in

the University, it was not unnatural that he should
take the same political views of the necessities of the
times, after the Duke of Wellington's declaration
against reform in any shape whatever, without really
deserving the imputation of self-interest. But even
supposing that self-interest had anything to do with
the change, it is quite certain that, when the Rubicon
was passed, he was a sincere Liberal, and that the
steps he took in a political direction, subject as they
undoubtedly were to misconstruction, were taken out
of a good conscience, and in the sincerest conviction
that they were a part of his duty." *

One of these steps was the prominent way in which
he came forward, on the occasion of the borough
election of 1835, to inform against some of the agents
on the Conservative side who had been guilty of
bribery.

In an "Address to the Reformers of the Town of
Cambridge," published by him the year previous, after
a contest for the town between Lord Monteagle (then
Mr. Spring Rice) and Sir E. Sugden, he had himself
said that, perhaps, there might be some among them
" who (did) not consider it becoming the character of
a clergyman to put himself very prominently forward
in election matters." And, " as a general principle,
he thoroughly coincided in such an opinion." He
proceeded, however, to state under what circumstances
he was induced to do so in that instance. He thought
there had been some want of fair play on the side of
the opposite party. He says, in justification of himself, " Independently of my desire to see the cause of

* The remark of a friend.

the Reform candidate triumphant, I clearly perceived
that a great 'moral struggle' had commenced, and
that if we were to prevail, it could only be by a
strenuous resistance to every kind of appeal that was
being made on the part of our opponents to the weak-
ness and to the corruptions of our human nature; and,
therefore, *as a clergyman*, I felt it a befitting occasion
for attempting to strengthen, both by advice and by
example, the weak or wavering minds of those whom I
knew to be with us at heart, but who were hesitating,
from some deficiency of moral courage, between obedi-
ence to the dictates of conscience and submission to
the powerful influence in operation against them."

On the above occasion his interference was restricted
to private reasoning with parties whom he considered
as having been unfairly influenced in the decision of
their votes, and to publicly putting forth his views on
the subject in the pamphlet referred to. But the year
following (1835) he took a much bolder step, though
entirely from conscientious motives. At this election,
the candidates were "Lord Monteagle, Professor
Pryme, and Lord Justice Knight Bruce—the contest
between Mr. Pryme (Liberal), and Sir J. L. Knight
Bruce (Conservative), being very severe." Bribery
was resorted to, and clearly proved against some of
the agents on the Conservative side. The Whigs
wanted to bring the offenders into court; but no one
would incur the odium of informing against them,
many, perhaps, declining under the apprehension it
might tell unfavourably on their interests in other
ways. Professor Henslow, though he did not join in
the squabbles of the town factions that arose out of

this circumstance, nor indeed take any part in the matter beyond allowing his name to be used in this way, boldly offered himself as the nominal prosecutor, not afraid to do anything he conceived to be for the public good, and willing to bear the brunt whatever it might be. The abuse and persecution which he had to sustain in consequence of this proceeding, is well known in Cambridge and has been often spoken of. Not only was the cry raised of " Henslow, common informer," whenever he appeared in the streets, but the same obnoxious words were placarded upon the walls in such large and enduring characters, that even to this day, * more than a quarter of a century after the transaction, they are still distinctly legible in some places. They were seen, and " smilingly pointed out to a friend," by the Professor himself within a year of his death, and I have, since his death, seen and read them myself. " His services were, however, deeply appreciated at the time, for he received three handsome testimonials: one from the town of Cambridge; another from the town committee for the suppression of corruption; and the third from a committee of noblemen and gentlemen."

The above circumstance deserves to be recorded mainly as exemplifying the moral courage which distinguished Professor Henslow, not merely in this instance, but in other instances also, as will appear afterwards. He had strongly advocated this virtue in his " Address to the Reformers " before quoted. He says, in recollection no doubt of his having had the courage to change his own politics a few years previous :

* Written in July, 1861.

" I would have every Tory consistent, and every
Radical consistent, and every Whig consistent, until
either of them shall have become convinced that he
has been in error, and then I would have him change
his politics, regardless of every risk, and despising the
shame which the world will heap upon him.   But
what I would have every man strive to possess is
' moral courage,' sufficient to declare his own opinions
unhesitatingly in the face of the world, and adequate
to maintain them unflinchingly against all influence
whatever.   .   .   .   A man of strict moral courage
will hazard everything rather than act against his conscience." He did not give this latter advice without
following it himself. Having once determined upon a
particular line of action, which he never did without
first maturely weighing it on all sides, and being
thoroughly satisfied that it was both right in itself,
and the right thing for him to do,—no fear of losing
popularity, no obstacles thrown in his way, no clamour
or even threats from without, could turn him from
his purpose. He went forward, utterly regardless of
what the world might think or say, and never de-
sisted till he had gained his point and carried out his
designs.

CHAPTER IV.

Some think that it was owing to Professor Henslow's zeal and activity in the above affair, and as a reward for his political services, that he received two years after from Lord Melbourne, the then Prime Minister, higher preferment in the Church. Others deny this, and allege that it was through the interest of Dr. Allen, Bishop of Ely, who had been formerly tutor to Lord Melbourne, and who, appreciating the talents and the character of the Professor, was anxious to obtain something for him that should not take him very far from the University, agreeably to the Professor's own wishes. Be this as it may, in 1837 he was presented by the Crown to the valuable living of Hitcham in Suffolk, worth upwards of £1000 per annum. He did not immediately cease to reside in Cambridge. During that and the following year he only spent the summer months at Hitcham; but in 1839, finding that the duties of so large a parish could not be properly attended to except by living constantly amongst his people, he came to the determination to quit the University altogether; and from that year to the period of his death the rectory-house at Hitcham was his sole residence, from which he was scarcely ever away, except when called by his professorial duties to lecture at the University.

The University had much cause to regret the removal of so distinguished and useful a member of its body. Independently of that generous and warm-hearted disposition, which had gained him so many friends, and which threw such a charm over his society to be enjoyed no longer, there was the withdrawal of that stimulus which he had so unceasingly given to the cultivation of the Natural Sciences. There was a manifest failing from that time of the interest felt in such subjects. The breaking-up of the Friday evening *soirées,* where all the resident naturalists had so habitually congregated, and which had been a source of so much enjoyment and instruction, told sadly in after-years upon the reduced numbers of young men who devoted themselves to Natural History pursuits. It was in order to supply the place of these evening meetings, that some of the scientific men who had attended them, and felt the blank occasioned by their discontinuance, instituted the Ray Club in the spring of 1837, "for the cultivation of Natural Science by means of friendly intercourse and mutual instruction." This Club, which is named after the celebrated naturalist, John Ray, formerly a Fellow of Trinity College, still continues. It consists of twelve resident members of the University of Cambridge, besides six associates chosen from the undergraduates and Bachelors of Arts, the meetings being held weekly at the rooms of the members in rotation. It was stated by the secretary in 1857 that "the members and associates have worked cordially together; and that the objects of the Club have been attained in as great a degree as its founders conceived to be possible." Still there were other influences besides those arising out of his evening meetings, exercised by Professor Henslow for the encouragement of Natural History in the University. There was the man himself, whose place no one could supply, with his engaging conversation, ever present and ready to say a word in behalf of his favourite science, and ever on the look-out for other minds which he might impress with a feeling of interest for it similar to his own. The altered circumstances, when he went away, appeared manifestly in the diminished number of students attending his botanical lectures. The falling-off commenced just about that time; and though, as before noticed, it was partly due to the additions made to the examination for a common degree, Professor Babington, his successor in the Botanical Chair, thinks that it was also due to the discontinuance of his evening parties, followed soon after by the discontinuance of his residence in the University.

But Professor Henslow himself only exchanged one field of usefulness and activity for another. The same energetic endeavours to do good, to direct the minds of the young especially, the same desire to make himself the friend of all, accompanied him into his new locality. He was now about to enter upon a sphere of life in many respects totally different from what he

had been previously used to. He had, indeed, been
accustomed to ministerial duties in Cambridge. But
his position there as curate in a town parish, where he
had simply to act under the direction of another, a
pious and exemplary clergyman, who at that time was
the incumbent, and always resident himself, was quite
distinct from that which he was now called upon to fill.
At Hitcham, he was about to take the sole charge of a
large and neglected rural district, in which it would be
necessary to exercise all his powers of judgment, skill,
and sagacity,—all his patience and perseverance, in
order to effect necessary reforms, to put things upon a
better footing, and raise the moral and physical as
well as the spiritual condition of its inhabitants. Professor Henslow was deeply alive to the responsibilities
of his new situation, and at once applied himself to
the work set him to do. He would seem to have
considered with himself in what way he could best
bring his energies to bear upon it, and to what extent
especially he could avail himself of those instruments
for good which he possessed in his scientific acquirements, in aid of the ordinary teaching which devolved
upon him as a Church minister.

Before stating in detail Professor Henslow's schemes
for the amelioration of his people, it will be proper to
say something of the parish of Hitcham itself, and its
inhabitants as he found them when he first went
there.

The village of Hitcham, which consists mainly of
one long straggling street, is situated on the road
from Hadleigh to Stowmarket, being distant from the
former town about six miles, from the latter town

seven. It is fourteen miles from Bury, and the same distance from Ipswich. The parish is said to have " contained, at the time of Professor Henslow's settlement in it, rather more than a thousand persons, scattered over an extent of 4117 acres of land. The real property in the parish was assessed at £6000 a year, yet there was only a dame-school in the village. The unemployed and vagabond labourers were so numerous that the poor rate in 1834 amounted to £1016, being an average sum of 27s. for every man, woman, and child in the village." Moreover, it is stated that "parish relief was not unfrequently levied by bands of forty or fifty able-bodied labourers, who intimidated the previous rector into instant compliance with their demands." With this state of things was associated, as may be imagined, a people sunk almost to the lowest depths of moral and physical debasement. Ignorance, crime, and vice appear to have been rife, even to a degree beyond what was too generally prevalent at that time among the Suffolk peasantry. The worst characters were addicted to poaching, sheep-stealing, drunkenness, and all kinds of immorality. Even witchcraft is said to have been practised by them. Others, if they did not join in such misdemeanors, were more fond of idleness than work, and lolled about on the banks and roadsides, dead to all sense of moral shame, so long as they could live at the parish expense, without any exertion on their part to maintain themselves and their families. Their houses are described as having been many of them little better than hovels, in which the common decencies of life could hardly be carried out. The Church was all but empty upon the

Sunday, and but little respect paid even to the marriage and baptismal rites. There had, indeed, been a resident rector before Professor Henslow, but he would seem to have been one of those easy-going men, of which there was formerly a large sample in the Church, who contented themselves for the most part with doing their Sunday duty, leaving the people to themselves during the week, and making no exertion to promote, in other ways than what the law obliged them to, either their temporal or their spiritual welfare.

Such was the moral waste which Professor Henslow was called upon to till and cultivate. Such were the evils he had to correct, such the people he had to deal with. The prospect would have disheartened many clergymen. His difficulties were the greater, as there were no influential persons in the parish to aid him in his attempts to correct what was wrong, and establish what was right. The farmers, who ought to have backed him up, were intellectually but little raised above the labourers they employed; and, with that obstinacy and prejudice which so often characterizes men of their class,—when they are much shut out from the world, and alike ignorant of and indifferent to what is done elsewhere for the amelioration of the poor,—instead of approving of the steps he took to raise the condition of his parishioners, doggedly opposed him in all his schemes, and threw every obstacle in his way they could. But the new rector was not a man to flinch or go back at the sight of difficulties. What he purposed, he was determined to accomplish, if possible. His aim was to recall his flock from sin and idleness to habits of soberness, honesty, and industry; to give them a love of independence, and to teach them to respect themselves, as well as those who were placed above them in life.

Moral and religious lessons would in the first instance have been utterly lost upon them. It would have been throwing pearls before swine. He wisely began with trying the expedient of winning them over by kindness and conciliation. He did what he could to amuse them. He got up a cricket club, and encouraged other manly games of a like character. He also brought into play some of the resources he possessed as a scientific man. He had an annual exhibition of fireworks, which his knowledge of chemistry enabled him to let off on the rectory lawn, and which have always a great attraction with the lower orders. On other occasions he would produce some of his natural and artificial curiosities, which, if they stared at in the first instance with that vacant look which betrays the ignorant and uneducated mind, they in time came to take more interest in, coupled with the desire of learning a little about their history and uses. No man knew better than Professor Henslow, how to adapt his language and illustrations to the capacities of his hearers, or how to gain the attention of those who were naturally least disposed to enter into his pursuits. If some of his parishioners took him, as was said, at first for a conjuror, in a few years they fully appreciated his talents and extensive knowledge, as well as the advantages they had themselves derived from his teaching.

But Professor Henslow passed on to other means of improvement. He lost no time in establishing a better school than had hitherto existed, aware, as all clergymen are who are solicitous about their flock, that to secure good wholesome teaching for the rising generation is one of the first things to be done in a parish where the adult population has been bred in ignorance and vice. It is unnecessary, perhaps, to state much on this head, as, with the exception of botanical lessons being after a while introduced into the school, as one efficient means of developing the faculties of the children, and getting them to take an interest in the works as well as in the Word of God— on which more will be said hereafter—there was nothing different in the mode of education in this school from what there is in the schools of other places. It may be just observed, that it was almost entirely due to his own liberality that Professor Henslow was able to put the school upon a better footing. He met with little support from any of his parishioners ; he had to bear the greater part of the expense himself in the erection of the school-house and the payment of a teacher, and it was only by very gradual steps that it was brought into that improved state in which it might be seen of late years, and which brought some individuals from a considerable distance to visit it.

To return to the labouring men of the village. Having in some measure gained their confidence, and won their hearts in the manner above alluded to, Professor Henslow next introduced ploughing-matches as a means of further encouragement to apply themselves to the work which their condition required of them, but which, hitherto, they had only lazily undertaken, or not undertaken at all. These matches were established in 1838, much to the satisfaction of the most deserving labourers, who entered into them with a good spirit, but not to the satisfaction of the farmers, who viewed them with an eye of jealousy when the prize fell to any of their neighbour's men, and not their own; and who interfered in such a way as "to render fair play impossible." They were consequently abandoned at the end of eleven years, and exchanged for the allotment system, a yet more efficacious means of raising the condition of the labourer, but which, from the fierce opposition raised by the farmers against this measure also, it was not till after long struggling and determined perseverance, that Professor Henslow succeeded in setting up.

Professor Henslow saw from the beginning that, in order to bring about any lasting improvement in the condition of the Hitcham labourer, it would be necessary to enlighten the employers as well as the employed, and to remind the farmers of the duties that devolved upon them as occupiers of the land. He has observed that, "All schemes, educational, recreational, or however tending to elevate (the labourer) in the social scale, are positively distasteful to some of the employers of labour, whom, nevertheless, we are bound to recognise as worthy men, not wilfully opposed to the comforts of those beneath them." * This, and more than this, was the case with the farmers of Hitcham. They held their men in grievous subjection. They viewed them as little better than slaves set to do their work, for whose temporal welfare they felt no concern. It is not surprising, therefore, that they should have left the rector to his own unaided

* *Gard. Chron.*, 1850, p. 691.

E

efforts in establishing a better school in the parish, while they set their faces altogether against the allotment system. Ignorance and prejudice have, indeed, often stood in the way of the latter measure. In many places the farmers are apt to think that the holding of an allotment will give the labourer a spirit of independence that will interfere with the service he owes his master. They grudge him a position, which, in their estimation, will bring him one step nearer to themselves, and weaken the authority they claim to exercise over him.

Yet it is these small holdings that, above everything, tend to strengthen instead of loosen, the bond of connection between man and master, and to make the former a more valuable and useful servant. The allotment finds the labourer employment in those hours which too many poor, for want of better occupation, spend at the alehouse, or fill up with vicious amusements. It makes him cheerful and contented. He has his own piece of land to fall back upon when other work is scarce, or not to be got. It brings him in also a small return, in addition to his wages, to be laid by against times of sickness and old age, when he is unable to work at all, instead of becoming chargeable to the parish. Nor is it true that it lessens his respect for his employer. It rather gives him an increased respect for both the farmer and the service he owes him, when intrusted with a small piece of land himself. He keeps a watchful eye over his own allotment, and it teaches him not to commit any breach of confidence in respect of the land he is set to cultivate for his master. Nor is it a boon only to the honest and industrious: many striking instances are on record in which "an allotment has been the means of reclaiming the criminal, reforming the dissolute, and of changing the whole moral character and conduct."

Perhaps every clergyman who has attempted to establish the allotment system in his parish, has met with opposition, more or less, at first. I myself experienced it when introducing it in my own parish in Cambridgeshire many years back. In my case, some said the men would give their masters short time and easy work, in order to reserve themselves for working on their own account at the end of the day. Others alleged that they would even steal their master's seed to sow upon their own land. A number of similar futile objections were raised, which in that instance the farmers soon discovered were quite groundless.

But in the case of the Hitcham farmers, it was a very long time before these were influenced by any reasoning or arguments that could be adduced for their conviction. Previous, however, to speaking of the battle which Professor Henslow had with them about the allotment system, it is right to state what he had endeavoured to do for their interests, as well as for the interests of the labourers in their employ.

He had, from the first, joined common cause with them as with others. He assisted them in their agricultural operations, which his extensive general knowledge enabled him to do, and taught them to farm on such scientific principles as would secure a more abundant return for their labours. He "had occasionally been present at the monthly meetings of the Hadleigh Farmers' Club, and had given them a few popular

lectures on scientific subjects connected with agriculture." After a discussion at one of these meetings,
on the "fermentation of manures," he was requested
to give them a lecture on that particular subject,
which he accordingly did, at the Anniversary of the
Club, on December 16th, 1842. The lecture was so
well received, that he was induced to follow up the
subject of the theory of manuring, in a series of
letters to the farmers of Suffolk in general, published
in the first instance, from time to time, in the *Bury
Post,* but afterwards collected into a volume, to which
was prefixed the above Address to the Hadleigh Club,
while a glossary of the more technical terms occurring
in the letters was given at the end.* Both the address
and the letters—which latter are fifteen in number,
dating from January 7th, 1843, to April 20th of the
same year—are admirably adapted, as well in style as
in matter, to the class of persons for whose guidance
they are intended. They convey scientific truth in
such a form, and in such language, as can hardly fail
to amuse as well as instruct. A friend seemed to have
rather taken the Professor to task for his " drollery,"
as the former termed it, thinking it, I suppose, not
quite suited to the dignity of the writer, or to that of
the subject on which he wrote. But Professor Henslow knew well the best way of " getting at " those
whose attention he wanted to secure, and that it would
not do to confine himself to dry technical details,
without now and then giving them a smack of the
humorous, if he wished to make them attractive to

* Letters to the Farmers of Suffolk, with a Glossary of
Terms used, &c. 8vo. London, 1843.

the particular class he addressed. In justification of himself on this point, he says, at the conclusion of his preface, " I must be content to lose caste in the eyes of the more discreet if I have been unfortunate in selecting an improper 'time to laugh;' but I shall not fear to give account of the few idle words in which I have indulged, seeing they were intended to serve a special purpose, which might not otherwise have been attained. In catering for the world's wide stage, it may not be amiss to remember that our audience are not all called to sittings in the dress circle; and a little sprinkling of broad farce may serve better to point a moral for some parties, than more sober comedy alone could have done."

In his address to the Hadleigh Club, he had proposed to the farmers to make experiments for themselves on the subject of manures, on a small scale, before venturing to operate on a large one in consequence of any suggestions he had offered. That they might be the better prepared for such an inquiry, he gave them a general view of the principles of nutrition in vegetables, upon which depends the right application of manures to increase the produce of the soil. He then treated of the relative value of the different kinds of organic and inorganic manures. He mentioned the desirableness of ascertaining " the extent to which farmers should allow the process of decomposition to be carried on in the dunghill, or whether they need allow the materials to ferment at all, before they are applied to the land." And he commented upon the bad practice of leaving dunghills by the road-side, and suffering all the liquid parts, the most valuable for the

crops, to run to waste. He told them " the preparation
of manure should be watched and attended to at home,
as carefully as the food prepared for cattle."

The main object of the letters, in pursuance of this
subject of manures, was to instruct the farmers how
to perform the experiments above recommended so as
to insure success in the results. He says, unless they
experimented for themselves, they might "be fifty
years in determining some point of importance which
might be settled in fifty months, or fifty days."

One experiment he strongly urged was to " test the
value of Liebig's suggestion, that gypsum should be
added to manure heaps to fix the ammonia."   He expressed a wish that " this experiment should be tried
by every farmer in Suffolk who felt any interest in the
progress of agriculture;" and with the view of getting
it tried by as many as possible, he was at great pains
to circulate widely in the county skeleton forms to be
filled up by the experimenters with the results at
which they severally arrived.

The number of farmers that responded to this call
was small at first, but gradually increased. Professor
Henslow, however, knew the difficulty of getting
practical men to deviate from the courses in which
they have long persisted, and was not discouraged at
what he had every reason to expect. He persevered,
though continually met by remarks that he was engaging in a hopeless task, in asking farmers to conduct such experimental inquiries as he urged. One
Suffolk agriculturist declared " that he was not
acquainted with more than one farmer whom he considered *qualified* for undertaking the experiment he

proposed." A scientific friend wrote to him, saying,
"As for hoping to get results *registered* by the
farmers, if you succeed, you will either prove yourself
a magician, or else that your Suffolk farmers are very
different from the Shropshire ones whom I have been
accustomed to." Others said he would not find fifty
persons in Suffolk (fifty was the number he wanted to
get) who had *sufficient public spirit* to co-operate in
the way he suggested. And after a certain number
had already come forward, it was insinuated that
though willing "not many would be *able* to fill up
the schedule in a satisfactory manner." One landed
proprietor out of Suffolk even declared "that he never
yet knew a farmer who could correctly fill up a
schedule of any sort." Some of the old farmers in
the neighbourhood, who had lived and prospered under
the system of farming into which they had been ini-
tiated half a century or more back, thought the Professor a meddling fellow, and laughed at the idea of his
being able to tell them anything worth their attending
to. This last was simply ignorance and prejudice
combined. But there was yet "another suggestion of
evil surmise," which Professor Henslow was more
careful to answer. It was feared in one quarter that if
"the experiment should *fail*, such a result would produce a bad effect, by shaking whatever confidence" the
farmers might before have been "disposed to place in
the recommendations of scientific men." "I hope it
will succeed," said a gentleman who first suggested
this possible contingency to the Professor. The Professor's reply to this was characteristic of the man, who
never was afraid to meet the truth in any shape.

" Really, Farmers of Suffolk," said he, " I am half
inclined to say that I hope it may not succeed, if I am
so little likely to be understood, that you suppose I have
formed any decided opinion about its success or failure.
If I had wished to play the conjuror, or had cared to
count upon your applause by foretelling success, I dare
say I could have contrived some experiment in which
success could have been secured, or at least not left
very doubtful.   Perhaps some of you might then have
been persuaded to consider me an agricultural Solo-
mon, and would have allowed me to dictate to you any
wild schemes or plans for agricultural improvements,
without any attention to common prudence or circum-
spection on your parts.   But I hate all such ma-
nœuverings and underplots, as I would detest the pious
lures and frauds put forth to prop or propagate an
unstable faith.   Look back through my letters, and
you will see that the object and end which I have proposed to myself has been the suggesting to you the
necessity of 'experimental co-operation;' and that
what I chiefly aim at in our present experiment is, to
show you how 'experimental co-operation' should be
carried on.   It would be very presumptuous and very
unphilosophical in me to say beforehand, that I am
positive of what will be the result of our experiment ;
but I do say that I am positive the experiment *cannot
fail* in producing some good result; and that is all
that a wise and cautious philosophy would have us
look to.   Let us first see what the result may be, and
then let us speculate afterwards upon what that result
may teach us.   With this determination, the experiment *cannot fail*—it must teach us something.   And

so of every experiment conducted on correct principles—it never fails. However it may fail in producing
that particular effect which the experimenter may
desire, or fancy he can secure, yet, even if he should
be completely baulked in his expectations, the experiment has assisted him in discovering *the truth*;
and it will, in consequence, teach him either to cleave to
the 'old ways,' or to turn to the 'new ways,' according as he finds that the one or the other practice will
best serve his purpose."

What the result proved, in the end, to be in the case
of the particular experiment here recommended to the
Suffolk farmers, is not mentioned in the "Letters."
It is stated, however, that no less than sixty-nine
farmers sent in their names as willing to conduct the
inquiry, and received schedules to be severally by them
filled up. This was a larger number than Professor
Henslow reckoned upon, and it showed that there was
at least more "public spirit" in the county than one
of his objectors anticipated. In some instances in his
own parish the Professor was himself present to super-
intend the necessary preparations for having the experiment properly carried out. He has given an
amusing account of the proceedings on one of these
occasions, and the whole testifies to the readiness with
which he would spare neither time nor trouble in
order to be useful to his neighbours, though in a
matter which in no way appertained to his immediate
profession as a clergyman.

And there can be no question as to the good that
resulted from thus stirring up the farmers to a more
methodical and scientific way of conducting their

E 3

agricultural operations. It helped to put them more upon a footing with the age. It served to open their eyes to the necessity of no longer persisting in old practices, if it could be shown experimentally that those practices were bad. Professor Henslow did not want to make them all philosophers; but he wanted to make them willing to co-operate in any experimental inquiries which scientific men suggested. Science and practice must go hand in hand: each must receive assistance from the other. The time was come when, as the Professor stated, " The *art* of husbandry should be converted into the *science* of agriculture." Science was extending its domain far and wide. Its influence was being daily more and more felt in the arts and manufactures. And it was now claiming to be attended to in places from which it had been hitherto shut out, and to have its voice listened to by those who had long turned a deaf ear to its suggestions. It was evident that if farmers continued to go on in the old jog-trot system, they would speedily be left behind in the race. All their operations are dependent for success upon the knowledge of certain first principles in chemistry and vegetable physiology, it might be added of geology also, without which they may, perhaps, secure a certain return, but can never get from the land all that the land is capable of yielding under a more judicious treatment, conducted by a better-educated class of men than farmers generally are.

There is reason to think, from the various communications addressed to Professor Henslow on the subject, that these letters had a wider influence than any he ever anticipated. By the circulation of the weekly press in which they were first published, their influence extended out of Suffolk, and attracted the attention of not a few proprietors and occupiers of land in other parts of England. Professor Henslow, however, did not carry on the subject further. As he reasonably observed, it would have taken up too much of his time, and called him too much away from higher and more important duties. He was content with having opened the way. He left it to others to propose other experiments, to be tried by the farmers. He himself " made his bow to them as a chemist," though he might still " gratify an itching for discussing some botanico-agricultural question or other," as he saw occasion. In his own humorous way of expressing it, he had concentrated his fire upon a single point, never slackening till a breach was made, but he could not pretend to maintain so rapid a discharge as before, after that the chief outwork had been gained.

There were other matters discussed in the above letters besides the management of manures. There was good advice given respecting the " stripping off the leaves of Plants," in reference to a practice adopted by some persons of pulling off all the leaves of their potato crops in order to increase the size of the tubers. Professor Henslow showed that the result of one experiment he mentions, rendered it highly *probable* that the practice was not advisable, though it might possibly be beneficial for certain purposes in the case of some other crops. He referred especially to the case of mangel-wurzel. In some places, he was informed, the leaves of this plant were " pulled two or three times during the period of its growth, whilst in other districts the practice was to pull them only a short time before the roots were dug up." He supposed this would not be done unless some advantage was perceived. Yet science and practice seemed here to be at variance. He took the question up in a botanical point of view. And he went on to state what the functions of the leaves are in connection with the nutrition of plants in general. He contradicted " the very commonly-admitted notion that the root *directly* nourishes the leaf, and not the leaf the root." Still he suggested that further experiments should be made with reference to the particular case of mangel-wurzel, and this was another instance in which farmers, if they wish to get at the real truth, should co-operate in carrying out an inquiry.*

The " Letters " conclude with a judicious appeal to the " country gentry " and the " rural clergy " to use their endeavours to promote a more inquiring spirit among the farmers in their respective neighbourhoods, and to secure their co-operation in taking such steps as may tend both to promote agriculture, and to raise the general intelligence of the agricultural community.

But there is one step towards bringing about this desirable end, to which Professor Henslow had already called the especial attention of his own farmers at the meeting of the Hadleigh Club; which forms the con-

* According to recent experiments by Professor Buckman, of Cirencester Agricultural College, it would appear that stripping the leaves off mangel-wurzel lessens the weight of the root crop by nearly one half.—See *Rep. Brit. Assoc.*, 1860, p. 38.

cluding part of his Address to that body, prefixed to
the volume of letters, and which, from its importance,
deserves to be recorded here. It respects the duty
which the farmer owes to the labourers in his employment. After speaking of manures, he adds,—

"Before I sit down I shall venture to say a few
words upon another subject on which I feel myself
much more qualified to give an opinion than upon how
your crops should be managed. There is a description
of culture which requires its special manure, and in
which I conceive you are as deeply interested as in any
which you carry on in the fields. You have the proper cultivation of your labourers to look to. This is
not the place, nor is this a befitting occasion for me to
appeal to you on any higher grounds than mere
worldly policy, for recommending attention to their
moral, intellectual, and social condition. One of the
best manures which you can provide for the description of culture I now allude to, is to secure your
labourers constant employment. I shall not enter
upon the wide field which this question embraces; but
I put it simply to you as a matter of worldly policy to
do so. I am no prophet; but it needs no prophet to
foreshow you what will certainly come to pass if your
labourers are thrown out of employ. If profits are to
depend in future upon increased produce, and not upon
high prices, then must there be an increase of general
intelligence among your labourers, to enable you to
take advantage of improved methods of culture; and
there must be increased labour, also, to carry them out.
I recommend to your serious attention that glorious
maxim of the wisest of earthly monarchs, 'There is

that scattereth and yet increaseth; and there is that withholdeth more than is meet, but it tendeth to poverty.'"—Prov. xi. 24.

These remarks were made when Professor Henslow was himself labouring to improve the miserable condition in which he found the Hitcham poor on his first arrival in the parish; and when, from what he saw immediately about him, he had reason to believe that the same poverty and misery prevailed elsewhere. The warning was much needed, and well-timed. There had, many years previously, been riots and insurrections throughout the kingdom among the labouring classes, who, in some cases, smarting under the real evils they endured, in others misled by artful and designing men, wickedly attempted to redress themselves in defiance of all law and authority. And within two years after the delivery of the above sentiments, there was a repetition of the same outbreaks, accompanied by crimes of the deepest dye. The villagers, from want of employment, and low wages, became discontented, and, blind to their own interests, at the same time harbouring a vindictive feeling towards their employers, they sought the destruction of the very property which could alone give them work. Scarce a night passed during the winter months, at one period, but the sky was lit up with incendiary fires, not merely in Suffolk, but in other of the Eastern counties. The alarm became general; and as the guilty parties showed little discrimination with respect to the objects of their attack, no farmer felt himself secure.

If the farmers were not themselves entirely to blame for all this, a great deal rested on their heads. Other causes may have combined to bring it about, but certainly things would never have come to such a pass if the labourer's position in society had been duly recognised by his employer — if his education had been more cared for, and no hindrances been thrown in the way of the advancement of his physical and moral condition. It must have been some consolation to Professor Henslow to think that he had spoken out on this subject on the occasion above referred to, and that he had himself done, and was still doing, so far as his influence extended, all he could to amend the evil. Now, however, that it was at its worst, the year 1844, he gave wider circulation to his opinions. He published, in the month of August in that year, a pamphlet, entitled, " Suggestions towards an Inquiry into the present Condition of the Labouring Population of Suffolk," besides which he made " a series of appeals to the landlords, the farmers, and the public at large," on the same subject, through the channel of the weekly press. With the view of raising the condition of the labourer, the great point he insisted upon was, that "justice should be done to the land; in other words, that no person should hold more land than he had capital fully to cultivate." He stated that, "if the land were well tilled, there would be no surplus of labourers, and, consequently, no want of employment; and that the possessors of the soil were bound to take care that its capabilities for employing and feeding the population should be duly called forth. As an illustration of the distress to which the labourers had been reduced by the contrary system, he entered into statements to show that in the parish of Hitcham not above three-fourths of the number of hands were employed that might be engaged on the land, and that the employment of an increased number would be to none so beneficial as to the occupiers themselves."

This advice, so freely tendered, was, no doubt, unpalatable to many; but it struck at the root of the evil, and it had its fruit in the end in his own neighbourhood at least, if not elsewhere. The Hitcham farmers seemed, after a time, in some measure awakened to a sense of their duty; a better spirit and feeling subsisted between them and the men in their employ, and, at the period of Professor Henslow's death, it is said that there was no redundancy of labourers in the parish.

But, before this improved state of things came about, he had succeeded in establishing the allotment system, to which, as before observed, there was a strong opposition raised on the part of the farmers, and to which subject it is now necessary to revert.

It was in 1845 that Professor Henslow first drew the attention of his parishioners, and the public generally in that part of the country, to the Allotment System, by the publication of " An Address to Landlords, on the Advantages to be expected from the general establishment of a Spade Tenantry from among the Labouring Classes." In this pamphlet he set forth the great improvement that would be likely to follow the introduction of the allotment system in the condition of those classes. He noticed that " a want of constant employment among a certain portion of the labourers was producing a vast amount of evil to the whole class." This evil he traced back to the fountainhead; "and it seemed to him morally certain that the labourers might be very easily restored to a state of comparative comfort and independence, provided all landlords were sufficiently attentive to the great 'moral duty' which attaches itself to the possessors of landed property." He had previously spoken on this subject in his address to the Hadleigh farmers, as also in his communications to the Suffolk papers before alluded to. But he here dwells upon it again, at the same time making extracts from other works, in which the above duty is recognized and clearly set forth. He wished these "two broad facts" to be kept in view,—"that the labourers (as a class) were not enjoying the hope and comfort they ought to possess in our free, favoured, and Christian country; and that this system of spade-culture had been shown to be capable of most mate-rially improving their condition." He thought there was sufficient evidence "in favour of the labourers being allowed some more direct interest in the soil than they at that time possessed;" adding, "Wherever the labourers are so completely prostrate at the feet of their masters, as they are in some of our villages, we shall look in vain for that wholesome independence of spirit which was so notoriously the attribute of our free peasantry in times not very long gone by." He observed, that some farmers were "not so sensible as they ought to be of the just claims of their labourers;" and that several, as was well known, actually took advantage of their helpless condition,—"becoming tyrannical and hard-hearted to an extent that could scarcely have been expected of professing Christians." He did not mean to consider the allotment system as a panacea for all the evils that oppressed the labouring classes, but it held out to them a very considerable resource, which might even be made something like an equivalent for the glaring insufficiencies of their system of servitude at that time.

Earnest as the above appeal was, and multiplied as were the arguments adduced by Professor Henslow in favour of the project he wished to carry out, they made but little impression upon the Hitcham farmers. With one or two exceptions, they continued to hold out for years in the most obstinate manner. Their selfish ideas still stuck to them: they would listen to nothing. Professor Henslow, however, did not go on waiting for ever for their consent before commencing the system he was so anxious to adopt. In an " Address to the Inhabitants of the Parish," circulated among them in Easter week, 1849—when some charity lands in the parish, which he was desirous should be laid out in allotments, were to be relet—he again drew their attention to the subject, and from that time took the matter seriously in hand. There were then but two allotments in the parish, which he had let out of his own glebe land. But so energetic were the measures he adopted to increase the number, and so popular did they become, that, in a similar "Address" to his parishioners the year following, he was able to announce the addition of "fifty others, of one quarter of an acre each," besides numerous applicants, still on the list, desirous of possessing an allotment when they could have one.

The farmers, finding the ground gradually giving

way beneath their feet, rose up, determined to defend
their own position to the last, and to make a firm stand
against the innovations thus introduced into the parish.
Thinking to crush the whole system, they assembled
in strong numbers at a vestry meeting held in March,
1852, and attempted to exact a pledge from each other,
which many gave, that they would "refuse all employment and show no favour to any day-labourer who
should hold an allotment." This resolution did not
at all disconcert the rector. The storm raged about
him, but he was proof against its violence. His moral
courage was not to be subdued by any such ungenerous
treatment of those whose cause he advocated. It
rather waxed the stronger, from the righteous indig-
nation felt at such conduct. He persevered, without
losing either his patience or his temper. He calmly,
but in as determined an attitude as that assumed by
the farmers, stuck up for the rights of the poor. He
declared himself to be their champion; and, in a short
printed statement which he circulated in the parish, he
conveyed to the farmers a severe and cutting rebuke for
their selfishness and illiberality. He told them plainly
that his mind was quite made up, that nothing
should deter him from carrying out his intentions,
and that he should treat with the utmost indifference
anything they said in dictation to himself, or in con-
demnation of the course he pursued.

This declaration seems to have had some effect upon
the Hitcham farmers, who, finding it was impossible
any longer to hold the ground for which they had so
long and so obstinately contended, at length withdrew
their opposition. They must, after a time, have seen

their mistake.   They must have been made sensible of
the marked improvement which took place in the
parish, and in the general conduct of the people, from
the period of the allotment system coming into operation, as I have been informed was decidedly the case.
At the time of Professor Henslow's death, the allotments amounted to nearly 150 in number.   The
scheme had been matured, and had brought forth its
fruits; and at the annual shows for the distribution of
prizes to the best and most industrious cultivators of
the soil, the Hitcham allottees are said to have often
" distanced all competitors in the excellence of their
produce."

Professor Henslow's moral courage has been already
alluded to in a former part of this memoir.   It was
quite as much called for, and quite as ready to show
itself, on the occasion just mentioned.   It is not every
clergyman that will boldly stand up against a body
of farmers, who have long had their own way in
everything, and who look upon the parson as one set
to attend to the souls of his people, but as having no
business to meddle in the general affairs of the parish.
It requires, indeed, some discretion and judgment to
know to what extent it is advisable to go, in the
attempt to carry out any particular measure, when it
causes an interruption of those amicable feelings that
should subsist between the clergyman and his parishioners.   Yet no one can otherwise than applaud the
course which Professor Henslow took in the matter of
the Hitcham allotments.   Whatever may have been
the case with his opponents, he himself was never put
out.   There was no breach of charity on his part.

He bore no ill-will to any one. He simply acted from a high sense of duty; firmly maintaining his ground, when a principle was attacked, but open to conciliation the moment his adversaries gave way.

It is necessary now to advert to some other schemes adopted by Professor Henslow for the good of his people. One of these was the institution of Horticultural Shows, at which there was a distribution of prizes, as alluded to above, when speaking of the allotment system, to the best and most industrious of the allotment tenants. These shows, of which there were two in the season, in the months of July and September, took place on the lawn before the rectory-house. They were first established in 1850, and there was a recurrence of them each year during the lifetime of the Professor. They acquired great celebrity, as well from the circumstance of the large numbers that flocked in from the surrounding neighbourhood to witness the proceedings of the day, in addition to the people of his own parish, as from the circumstance of the detailed notices of each meeting which generally appeared in the county papers the week afterwards. The high opinion that was entertained of their utility, was shown by the influence they had in causing similar shows to be got up in other places; one especially that was established at Thornham Hall, Suffolk, by Lord Henniker. Besides prizes for wheat, there were prizes distributed for fruit, flowers, vegetables, and honey. There were also sometimes prizes offered for works of mechanical ingenuity, such as figures of animals, &c., carved in wood. This was done with the view of encouraging the labourers to

spend their long winter evenings profitably, and to
find them an amusing occupation. Tents were pitched,
that for receiving the productions of the cottagers'
gardens being fifty or sixty yards long; and, on one
occasion, "there were no fewer than ninety-seven ex-
hibitors." A spectator, who went to visit the show
held Sept. 28, 1853, speaking of the exhibition of
vegetables and flowers by the villagers, says,—" It did
great credit to the competitors, displaying an amount
of skill in cultivation, which few who did not see it
would believe could exist in a remote country village." *

Besides the tents for the more special purposes of
the show, there was always one tent allotted to a miscellaneous collection of specimens in natural history,—
animals, birds, nests, reptiles, insects, &c.,—in addition
to specimens in the arts, some from foreign countries,
to illustrate the domestic furniture, clothing, and
utensils of other nations,—models also, and antiquities.
This the Professor called his "Marquee Museum," the
various contents of which were got together at the
cost of no little trouble, to attract the attention of
visitors, and to give an additional interest to the meeting. On one occasion the dimensions of the trunk of
the great mammoth tree (Wellingtonia) were traced
out on the lawn, with a diagram showing its relative
size in comparison with other trees. On another occasion there was to be seen at the top of the tent for the
villagers' productions, "a most elaborate model of a
gentleman's house and garden, with flower-beds, walks,
shrubberies, fountains and visitors, all carefully and

* *Gard. Chron.*, 1853, p. 646.

beautifully modelled." But it was not merely to gratify the eye that he catered for the company in this
way. At intervals, to relieve the more routine business
of the day, he would summon to the Marquee
Museum as many as liked to come, and, for a few
minutes at a time, deliver those little lectures,—" lecturets " he termed them,—which so peculiarly distinguished these shows, and which none but himself
could have so felicitously adapted to the mixed audience
before him. He knew exactly the kind of things that
would interest his hearers, and the kind of language
to use whilst explaining them ; and he had something
ready for each. He would talk to the women about
the Manchester cottons, or tell them the way in which
they washed linen or conducted other domestic operations in some places abroad. He would explain to
them the action of the common pump, or the way in
which some of their common utensils were made, or
show them different articles and curiosities, worked up
for use or ornament by savage nations. The village
blacksmith might learn something about ironworks ;
the glazier see different kinds of glass, and hear the
several processes by which they were obtained. And
all had their ideas raised above the productions of
their own neighbourhood, by a sight of foreign plants,
birds, and other animals. Sometimes the Professor
would produce living specimens of toads and snakes,
and teach his people not to repute as venomous, according to the common superstition, creatures which were
perfectly harmless, and which it was wanton cruelty to
destroy. The farmers he would lecture on the weeds

of agriculture; or show them mummy wheat, rice in the ear, or samples of corn from the colonies, or magnified representations of the funguses which produce the diseases in wheat; at another time he would teach them not to trust to the moon when haymaking, for a change of weather, and try to remove other hindrances to the spread of knowledge and truth, arising from vulgar errors and vulgar prejudices.

The exhibition of these different articles, and this familiar way of addressing the people, served not merely to enliven the day, but to cement the bond of union between the people and their instructor. It led them to regard him as their friend and companion, as well as their minister; and it gained for him a more attentive ear, when at other times he had occasion to inculcate upon them lessons of a higher and more important character.

Nor should it be omitted to state that, for the amusement of the lads and young folk who came to the shows, but to whom the holiday on the occasion was the chief attraction, he was good-natured enough to set up swings and poles for gymnastic exercises, and allow them to play foot-ball and other games on the lawn. In short, the whole was a scene of festivity and mirth, harmony and goodwill, instruction and entertainment combined, such as, it may be safely said, has never been surpassed, if ever witnessed, in the garden of the village clergyman, and such as the inhabitants of Hitcham will long remember. All which was farther promoted by a plentiful supply of tea and other good things to the villagers in the evening, winding-up with " God save the Queen " as a *finale* to the whole.*

As a specimen of one of these days, I have annexed a programme of the Horticultural Show held on October 3rd, 1860, being the last that ever took place.

HITCHAM LABOURERS' AND MECHANICS' HORTICULTURAL SOCIETY. SECOND SHOW, OCT. 3RD, 1860.

———

" 8 to 10 A.M. Specimens received on the rectory lawn.

" 12. Marquee museum ready for inspection. Lecturets as opportunity offers later in the day. Among additions since the July Show, observe

" 1. Five unpublished lithographic portraits of their Royal Highnesses, the Princesses Alice, 1859; Helena, 1849; Louise, 1851; and the Princes Arthur, 1859; Leopold, 1859.

" 2. Photographs of highly-magnified objects of Natural History; among which are a fly's tongue and eye; Parasites (lice) of Man and Ox; Spider's claws; Saws of a Saw fly; fossil Foraminifera. N.B. The

* Several interesting accounts of these Horticultural Shows have been published by different persons who have visited them, in the *Gardener's Chronicle;* see vols. 1854, p. 629; 1856, p. 646; 1857, p. 679; and 1859, p. 623.

Perhaps the most detailed, as well as the most amusing, will be found in the *Literary Gazette* of July 9, 1853, from the pen of Mr. Lovell Reeve.

F

Chalk (in round numbers, 1000 miles in length, 500 in breadth, and 1000 feet thick,) has been mainly produced by these microscopic creatures! Section of Pine-wood with circular disks on the cells.

" 3. Magnified representations of 151 forms of Snow Crystals.

" 4. Case containing living specimens of our smallest British Mammal, the Harvest Mouse.

" 5. Pearls from four British Mollusks, viz., Oyster, Periwinkle, Pearl-muscle, Fresh-water Clam.

" 6. For the sake of Children, young and old. A case containing a " Heap " of Shells and Corals; also, under a Glass shade, a new Device from the last Horti-cultural Show in " Fairy-land."

" 1 P.M. Show-booth ready for inspection.

" 2½. Prizes to Village Botanists of the first class, for (1) Wild-fruit Posies; (2) Dried-grass Posies. N.B. The species to be named in each case. (3) Herbarium specimens. School Report.

" 3. Allotment Report. Prizes distributed for (1) Superior culture; (2) Hatcher Sweep-stakes; (3) Specimens exhibited at this Show.

" Thanks to the Judges, Donors of Prizes, and all others aiding and abetting our pro-ceedings.

" 4. Begin to remove Specimens from the Booth.
Not to forget to restore Cheque and Prize
Tickets to the Stewards.

" 5. Ready for Tea. Ticketless Babies (0 to 2
years old) and Ticketed Ditto (2 to 4 years)
admitted as heretofore.

" 6. God save the Queen. Good Night. May
the occasion prove a blessing, without a
single instance of offence against the laws
of God or Man."

It will be observed in this programme, that at 2½ P.M.
prizes are distributed to the village Botanists. This
leads to mentioning the part which the school-children
took in these meetings, in connection with the subject
of Botany, as taught by Professor Henslow in the
Village School.

Without omitting other branches of knowledge,
some no doubt of more importance, he thought that
Botany might, to a certain extent, be conveniently
employed "for strengthening the observant faculties,
and expanding the reasoning powers, of children in all
classes of society." Independently of the value of
botanical knowledge abstractedly considered, the study
of it leads to further advantages, and may be service-
able in many ways. It gives children a habit of observing nature, teaches them what kind of facts to
notice, and how to observe correctly, so as to render
their observations of avail to themselves or others.
Even in the case of the children of the lower orders,
it tends to make them more useful in the several
callings they are likely to exercise in after-life. Young
women in service, who often have the care of the

children of the rich,—lads employed either in the farm
or garden,—still more, those who may be engaged as
pupil-teachers in other schools,—all these have the
opportunity, more or less, of turning such knowledge
to account. It furnishes them also with innocent and
rational amusement in those leisure hours, which so
many servants and poor idly throw away when their
required work is done. Above all, it tends to raise
their thoughts to the contemplation of the Creator,
and to make them mindful, as well as observant, of
that infinite wisdom and goodness, of which they see
everywhere around them such abundant proofs.

But in order to obtain these beneficial results,
botanical lessons must not be confined to telling the
children the names and properties of plants, or how
they may be artificially grouped, but must be directed
to teaching them their *structure,* and their true affinities as dependent upon that structure.   This was what
Professor Henslow strongly insisted upon, and made
the groundwork of the lessons given in his school.
Nor can it be effectually carried out without employing
" certain technical expressions," which alone convey
" scientifically accurate ideas."   Accordingly, his first
step was to get the children thoroughly to master these
necessary terms, and to understand their meaning.
His habit was to attend the school regularly every
Monday afternoon, for the purpose of giving a lesson
in Botany, from an hour and a half to two hours in
length.   The botanical pupils were all volunteers, and
limited in number to forty-two. They varied in age
from eight to eighteen, and mostly entered with great
spirit into the work set them, seeming thoroughly to

Children wishing to learn Botany will be placed in the Third Class, when they shall have learnt to spell correctly the following words :—

CLASS.	DIVISION.	SECTION.
(I. *Exercise*.)	(II. *Exercise*.)	(IV. *Exercise*.)
	⎧ 1. Angiospermous.	⎧ 1. Thalamifloral.
	⎪	⎪ 2. Calycifloral.
1. Dicotyledons.	⎨	⎨ 3. Corollifloral.
	⎪	⎪ (V. *Exercise*.)
	⎩ 2. Gymnospermous.	⎩ 4. Incomplete.
	(III. *Exercise*.)	
	⎧ 1. Petaloid.	⎰ 1. Superior.
2. Monocotyledons.	⎨	⎱ 2. Inferior.
	⎩ 2. Glumaceous.	
3. Acotyledons.		

Children in the Third Class, who have learnt how to fill in the first column of the Floral Schedule, and to spell correctly the following words, will be raised to the Second Class :—

Pistils and ⎱ of Ovary (with Ovules), Style, and Stigma.
Carpels ⎰
Stamens, of Filament and Anther (with Pollen).
Corolla, of Petals ⎱ or Perianth, of Leaves.
Calyx, of Sepals ⎰

Children of the Second Class who have learnt how to fill in the second column of the Floral Schedule, and to spell correctly the following words, will be raised to the First Class :—

C. Mono-di-,&c.,to poly- phyllous, -sepalous, -petalous, -gynous. V. Mon-di-, &c., to poly- androus, -adelphous. Di-,tetra-,dynamous. Syngenesious.	V.	C.	V.	C.	V.	C.
	0. An-	A-	5. Pent-		10. Dec-	⎫
	1. Mon-	o-	6. Hex-		11. Endec-	⎬ a-
	2. Di-	—	7. Hept-	⎱ a-	12. Dodec-	⎭
	3. Tri-	—	8. Oct-	⎰	20. Icos-	
	4. Tetr-	a-	9. Enne-		∞ . Poly-	—

Children of the First Class will learn to fill in the third column of the Floral Schedule, and to spell correctly the following words :—

Hypogynous.   Perigynous.   Epigynous.
Epipetalous.   Gynandrous.

Monday Botanical Lessons at 3 P.M., at the School, to include,

1st.—Inspection of a few species, consecutively, in the order on the plant-list. Anything of interest in their structure or properties will then be noticed.

2nd.—Hard word exercises. Two or three words named one Monday are to be correctly spelt the next Monday.

3rd.—Specimens examined, and the parts of the flower laid in regular order upon the dissecting-boards. The Floral Schedule to be traced upon the slates, and filled up as far as possible. Marks to be allowed according to the following scale :—

OOL BOTANY.

	No.	Cohesion, Proportion.	Adhesion (Insertion).	Classifica-tion.
P. C.	1/3	a-, mono-, &c., gynous 2	Superior or Inferior 2	
St.	1	an-, mon-, &c., androus 2	Hypo-, &c., gynous 4	Class 1
f. -	1	{ mon-, &c., adelphous 3 { Di-, tetra-, dynamous 3	Epipetalous 4	Division 2 Section 3
a. -	1	Syngenesious 2	Gynandrous 3	Order 4 Genus 3
{ C. P. { C. S. or P. L.	1 1 1	a-, mono-, &c., petalous 2 a-, mono-, &c., sepalous 2 a-, mono-, &c., phyllous 2	Hypo-, &c., gynous 4) Inferior or Supe-) rior 2	Species 2

4th.—Questions respecting Root ; Stems and Buds ; Leaf and Stipules ; Inflorescence and Bracts ; Flower and Ovules ; Fruit ; Seed and Embryo.

Regulations respecting Botanical Prizes and Excursions.

Prizes awarded according to the joint number of marks obtained at Monday Lessons, from Schedule Labels filled in at home, and for species first found in flower during the season.

Botanical Excursions attended only by those who obtain a sufficient number of marks at Monday Lessons. Two Pic-nic Excursions during the summer, within the precincts of the parish, open to children in each of the three Classes. Other Excursions within the parish are open only to those of the Second and First Classes. An Excursion to a distance from the parish for those of the First Class only who obtain the requisite marks.

The First Class may attend (at the proper season) at the Rectory on Sundays, after Divine Service in the afternoon. Objects of Natural History, in the Animal, Vegetable, and Mineral Kingdoms, will then be exhibited, and such accounts given of them as may tend to improve our means of better appreciating the wisdom, power, and goodness of the Creator.

A copy of the above scheme is given to every child, however young, who is ambitious of being classed as a volunteer Botanist.

Example of a Floral Schedule filled up.

				CL. Dicotyledons.
P. C.	1/2	Monogynous.	Superior.	Div. Angiospermous.
St.	6	Tetradynamous. Hexandrous.	Hypogynous.	Sec. Thalamifloral.
C. P.	4	Tetrapetalous.	Hypogynous.	Ord. Brassicanths.
C. S.	4	Tetrasepalous.	Inferior.	Gen. Wallflower.
Harriet Sewell, No. 7				Sp. Common.

enjoy it. They were divided into three classes. A certain number of words, however, expressive of the characters of some of the leading divisions under which plants are arranged, were given them to spell correctly, before they were allowed to enter even the lowest of these classes. When they had gained their place in the third class, other words, designating the different floral organs of plants, were given them to spell in like manner. There was also put into their hands what was called the " Floral Schedule," a portion of which they were required to fill up. When able to do both these things correctly, they were raised to the second class. Higher lessons of a similar character were then set them, which they were equally to master before being raised to the first. All this, however, will be rendered much more intelligible by an inspection of " The Printed Scheme for Monday Lessons," a copy of which is given on the annexed Table. A copy of the same scheme was " given to every child, however young, who was ambitious of being classed as a volunteer botanist."

It will be seen that the first thing mentioned in the Monday Lesson is the " inspection of a few species, consecutively, in the order on the plant-list." This refers to a printed list, drawn up by Professor Henslow, of all the plants growing wild in the parish of Hitcham, with the addition of a few common trees in plantations. The numbering of the orders, genera, and species in this list agrees with "Hooker and Arnott's British Flora," the orders having " been *Anglicized* by changing the terminations of the genitive cases of their typical genera into ' anths ' (*flowers*) ; as Ranuncul-i, Ranunculanths, &c." A copy of this Plant-list was

" given to every child who had fought its way into the
third class, and could write down from memory the
thirteen words of the five exercises at the top of the
printed scheme."

Before speaking of the way in which this plant-list
is used, it is necessary to refer to some of the apparatus with which the Hitcham School was furnished.
There was what was called the " plant-stand and
labels." This consisted of a framework, cheaply constructed of eight thin strips of deal, open in front
and behind, and closed at the ends, each of the end
pieces being three inches long and two inches wide.
" The top is perforated to receive eighteen common
cast phials, one inch diameter. Sixteen plant-stands
will carry half a gross of phials, rather more than the
number of wild-flowers likely to be obtained in blossom
at the same time." The stands are furnished in front
with " slit-strips," at the top and bottom, for the
purpose of introducing the labels to be affixed to the
plants. These labels are made of thin card-board,
having the general terms—class, division, section,
order, genus, species, &c.—printed upon them, against
each of which terms, the particular class, &c., as the
case may be, is to be inserted in manuscript by the
children. " The labels, which are not in immediate
use, are methodically arranged in a box, and the
numbers, indicating the order, genus, and species (as
in the plant-list), are written at the back, near the
top of each. If the labels are put away in an upright
position in the box, they can very readily be arranged
and referred to by these numbers."

" The children are required to bring fresh specimens

of all the wild-flowers they can procure, in order to keep up the display; but only those who are well acquainted with their names are authorized to place them in their proper phials."

Next, there were the "dissecting-boards," made of thin deal, twelve inches long and nine wide, one of which was supplied to each child. "Across the upper half of each board is pasted a paper with four compartments. Opposite these, the names of the four floral whorls, and their subordinate parts, are printed (the whole might be in manuscript), together with the adjective terminations for expressing botanically the numerical and other relations between them, noticed in the floral schedule" in the printed scheme.

"The dissecting-boards are numbered to correspond with numbers assigned to the children. This affords a ready method of referring from their exercises to the lists in which their merit marks are registered. To prevent copying, and to compel the children to depend upon their own observations, a distinct position is assigned to each in the school-room; and those of different classes are so intermixed, that no two of the first and second class sit contiguous to each other.

"A flower is handed to each child: a more difficult specimen being usually given to those of the first and second classes than to those of the third. The flowers are pulled to pieces, and the separate parts of each 'floral whorl' are placed in the appropriate compartments on the dissecting-board. A single glance shows whether this preliminary step has been properly performed. When these pullings to pieces (dignified by the name of dissections) have been completed, and the children

have determined the number, cohesion, and insertion
of the parts in the floral whorls, they next record upon
their slates the results of their observations, and the
inferences they have drawn from them."

The scheme given in the Table above, " will show
how much of this schedule is expected of children in
the second and third classes. The children rule their
slates for this exercise before they begin to dissect, the
elder often assisting the younger. As they severally
complete it, they deposit their slates in one part of the
school, and walk into a contiguous class-room, fitted
with the usual school-gallery, one bench above another.
The slates are conveyed to the rectory, and are inspected the same evening. Errors are corrected, marks
assigned for merit, and a few comments made upon
the slates in some cases. The slates are returned to
the school on Tuesday morning, when the school-
mistress, or the pupil-teacher, reads aloud the remarks."

We come now to speak of the " plant-list " exercise.
The class-room in which the children take their seats,
after having filled up the " floral schedule " as far as
possible, is fitted up with the plant-stands. It is
further furnished with two or more " suspending-rods
of deal, six feet long, one inch broad, and half-inch
thick, for exhibiting dried specimens from the school
herbarium, diagrams, plates, and other illustrations, as
occasion may require."

" A few specimens from the herbarium are exhibited. These are selected in regular sequence. Where
an order is so extensive that all the species cannot be
exhibited at one lesson, a few specimens are brought

forward at one time, whilst progress is made with the orders next in the plant-list. It is advisable to confine a lesson to the details of a single order or two, even though each may contain not more than one local species.

"The children are now told to look at the plant-list, and a copy of it is distributed to every two or three for the occasion. They place their fingers upon the order that was noticed the previous Monday. They then pass on to the next order, and repeat its name aloud, and are told to remember to learn how to spell it correctly before the following Monday, when they have to write it (together with the floral schedule exercise) on their slates. If the order is correctly spelt, one mark is allowed for this part of their performance; but a mark is taken away if no attempt has been made approaching to accuracy. Attention is then called to the few (say half-a-dozen) dried specimens from the herbarium suspended before them, and all who know their names repeat them aloud. Looking to the plant-list in their hands, they state how many genera and species belong to the order about to be noticed. This kind of discipline keeps up their attention, and assists greatly in familiarizing them with the names of plants and their classification.

"The more prominent characteristics of the order, the more striking properties of some of its species, whether British or exotic, and anything interesting or useful, are then exhibited or mentioned. A few questions, likely to elicit more or less pertinent answers, are interlaced, and, for the most part, inquisitive eyes and smiling lips prove the time has not been wasted."

The above relates to the botanical lessons taught in the school. But, besides these, there were the "home exercises," required only from those who chose to undertake them. These exercises consist of the filling up, out of school hours, what are called the "label-schedules," in which are put down, according to a printed blank form, the class, order, &c., of the several plants the children collect, together with the number of each of the floral organs, their cohesions, insertion, &c., more on this head being required from the elder children than from the younger. The forms are somewhat similar to those employed for the floral schedule in the scheme. A portion of the plant itself is sent in with the exercise, to show that the plant has been really obtained, and marks are allowed, more or fewer, in proportion to the pains taken to have the exercises correct.

I have been the more particular in describing Professor Henslow's method of teaching botany in the Hitcham School (and as much as possible in his own language), from the notoriety it has gained, "and the success that has attended it as an educational measure." It has been taken up by the Committee of Council on Education, and botany is now taught in some other schools in different parts of the country. At Marlborough College especially, as I am informed, the introduction of botany has been attended with great success. Botanical prizes were instituted here in May, 1860, by one of the masters, who applied to Professor Henslow for advice on the subject, and acted principally upon his suggestions. About eight scholars went in for the prizes, and the collections of plants

made by the successful candidates were very creditable
Professor Henslow was to have assisted in forming
some botanical classes, but his death prevented this.
He was also to have attended at the college, in order
to examine the collections of the candidates—the rules
to be observed by them having been chiefly drawn up
by himself—but for the same circumstance.

Those who wish for further information on the subject of teaching botany in schools, should consult a
little tract, published by Professor Henslow in 1858,
entitled "Illustrations to be employed in Practical
Lessons on Botany; adapted to beginners of all
classes." All the passages above marked in inverted
commas are taken from this tract, where there will be
also found figures, with more full explanations than
I have given, of the different kinds of apparatus used
in the school. The tract was prepared for the South
Kensington Museum, and specimens themselves of the
apparatus may be seen in that Institution.

He also issued, for the Committee of Council on
Education, Department of Science and Art, "A Series
of nine Botanical Diagrams," intended chiefly for
school-room purposes, for which they are admirably
suited. They are of a large size, and coloured; and
for "scientific accuracy, and general artistic effect,"
can hardly be surpassed. They serve to "teach beginners how certain important technical terms are employed. The illustrations repeat, as much as possible,
the application of the same terms to parts exhibited
under those different forms and associations by which
plants are either naturally or artificially grouped into
large classes, and their subordinate divisions and sections."

He was engaged, at the time of his death, on yet
another work on School Botany, the title of which was
to have been "Practical Lessons in Systematic and
Economic Botany, Educational and Instructional, for
the use of Beginners in Village Schools and upwards."
This little work had been long promised, and was left
in a forward state, but not ready for publication.

The extent to which many of the village children in
Hitcham, especially the elder ones, became acquainted
with botany, the readiness with which they could name
most of the wild flowers of the neighbourhood, referring them to their right places in the system, the
leading divisions of which they had thoroughly mastered, the correctness with which they could distinguish and describe the several parts of each plant—
while it astonished visitors to the school—afforded a
proof of Professor Henslow's skill in imparting scientific knowledge to minds that might be thought by
some incapable of receiving it. It was, indeed, by the
help of his school children, whom he had thus trained
to habits of observing, and who not unfrequently
brought him species that he himself had never found
in the neighbourhood, that he was able considerably
to enlarge his catalogue of Hitcham plants.

The children had been further taught to dry the
plants which they found, and each to form for himself
or herself a small herbarium. Mention has been made
above of the school herbarium (much of which was
got together by the children), when speaking of the
plant-list exercise; and in the "Illustrations" tract
many practical directions and useful hints are given for
the help and guidance of children in this operation,
with figures of the "collecting portfolio,"—"packet of

drying papers," containing the "plants under pressure,"—the plant "ventilator," to allow of the air passing in sundry places through the packet,—and, lastly, the "deal cabinet," of a cheap and simple construction, to hold the herbarium when formed.

Allusion has been made to the part which the children took in the horticultural shows described above. This arose out of their botanical acquirements. They were invited to bring to the shows—in addition to a selection of dried plants, prepared as instructed—parcels of fresh wild-flowers, consisting of such different species as the season and the neighbourhood produced. These were to be put neatly together in nosegays, all being properly named; and prizes were awarded to those children who displayed most skill and taste in this operation, and who were most correct in assigning the right name to each plant. At the July show, in 1858, no less than fifty-six school children competed for "wild-flower nosegay" prizes, and twenty-six received prizes. Some of the herbaria, too, exhibited by the children at these shows, were surprising. One pupil-teacher is said, on one occasion, to have "actually collected in rural strolls, and afterwards dried and correctly named, more than 250 specimens of plants."

That the children took an eager delight in their botanical lessons could not be more strongly shown than by their grief expressed on the occurrence of anything to cause an interruption of the lessons. Professor Henslow, in one of his annual addresses to his parishioners, speaking of the school, says, "No one who had heard the lamentations uttered upon my

announcing, at our last lesson before Easter, the necessity of six weeks' absence at Cambridge duties, could
possibly have doubted the great interest the children
take in these exercises."

I am also glad to be able to give the opinion of one
of Her Majesty's Assistant Inspectors of Schools, who
twice inspected the school at Hitcham, as to the effect
produced upon the minds of the children by lessons
in botany. He says, in a private communication to
myself,—

"That the botanical lessons, as handled by the Professor in his own National School, did draw largely
upon the intelligent powers of his little pupils' minds
there can be no question. The simple system to which
he had reduced his plan of making the children break
up the various specimens into their component parts,
arrange those parts, observe their characters and relations to each other, and thence *arrive at conclusions for
themselves,* was very far from being the mechanical
process which many, before witnessing it, might have
supposed 'botany in National Schools' to represent.
And I think it not at all unfair to say, that these
children, who, out of school, were (as I had many opportunities of judging) much more conversable than
the generality of children in rural parishes, owed a
considerable share of the general development of their
minds to the botanical lessons and the self-exercise
connected with them."

Professor Henslow had a playful way with children,
which won their affections as well as their attention to
what he was teaching them, and which was one secret
of his success. He would always speak kindly to them,

and encourage them in different little ways. All who
competed for the wild-flower nosegay prizes, though
they did not succeed in getting a prize, were allowed
a pinch of "white snuff," as he jokingly called it—or
sugar-plums. He generally had a snuff-box full of
these sugar-plums in his pocket when he went into
the village, offering a pinch to any of the little children whom he happened to meet.

A feeling of honest pride, if successful in obtaining
a prize on so public an occasion as that of the Horticultural Shows, stimulated the children to make
further progress in the science they had been taught,
while it encouraged others to commence as students,
in the hope of being one day able to deserve the like
distinction at the hand of the Professor.

Another encouragement was the desire to be allowed
to join in the botanical excursions undertaken occasionally in company with their instructor. These days
were very much relished and looked forward to. It
will be seen, however, from the printed scheme, that
they were open only to those children "who obtained
a sufficient number of marks at Monday lessons."
Some of these pic-nic excursions were open only to
the first and second classes, while the more distant ones,
out of the parish, were reserved for children "of the
first class only, who obtained the requisite marks."

On one occasion, by invitation from Professor Henslow's old college friend, the Rev. H. Kirby, a dozen
of the Hitcham village botanists "enjoyed a pleasant
excursion to Great Waldingfield, where they met a
dozen of his village school-children, and the joint
party proceeded to ransack the neighbourhood for

plants." Several were found which do not grow at
Hitcham, and they were duly preserved and exhibited
at the next show.

On another occasion Professor Henslow made an
educational trip to London with a small party, consisting of himself and one of his daughters, the school-
mistress, two pupil teachers, and two or three of the
elder girls from the first class of his Hitcham National
School. I cannot do better than give some particulars
of this trip as supplied to me by the same School
Inspector before alluded to, who met the party in
London, and who saw something of their proceedings.
The trip took place on Monday the 15th of November,
1858, and the party did not return till the Saturday
afterwards.

" During the week " (the Inspector writes word) " the
party visited, I was told, Kew Gardens, the Crystal
Palace, and I believe the South Kensington Museum
and British Museum. On Thursday in that week they
met me by appointment at Clapton, in the girl's school
of the London Orphan Asylum. Professor Henslow
had applied to me to name a school, easy of access, in
which they could see carried out into practice, and in
a high degree of perfection, methods of discipline
enforced by moral power, and accompanied with such
standards of attainment in the children as would fall
within the compass of their own judgments to form an
estimate of. This was his idea of meeting, in a practical manner, suggestions of improvement in his own
school-teaching power, which had been elicited and
canvassed between us at my recent visit of inspection
at Hitcham. In return for the profit thus to be made

of this school, the Professor undertook to give a lecture in the evening to the orphan children and the friends of the Institution in their school-room. The result was a most interesting lecture, which brought quite within the limits of their childish comprehensions a subject he entitled, 'Vegetables the ultimate source of Food to Man.'   He was, too, at the pains to bring with him a large box of specimens with which to illustrate his lecture.   His visit is still kept in grateful remembrance at that excellent Institution, in which everything is both done and received in a pure spirit of kindly affection."

The excursions which Professor Henslow made with his little class of village botanists, were not the only ones he organized in his parish.   There were others on a larger scale at certain times, open to all his parishioners rich and poor, all well-conducted poor at least, who liked, and who had it in their power to join.

It was in 1848 that the excursions now spoken of were first planned; and the success that attended them, and the enjoyment they afforded, had a powerful effect in strengthening the attachment between minister and people, as well as enlightening the minds of the class of poor for whose especial benefit they were set on foot.

When Professor Henslow first came to reside at Hitcham, it struck him " as a bad feature in the habits of the place, that the labouring population had few opportunities allowed them for enjoying a day's holiday beyond the one customary holiday at Whitsuntide. Plenty of hard work, without recreation for body or mind, was the usual life of the best labourer, and the

more especially, if he were one of that description of
men who ' do their duty in that state of life to which
it has pleased God to call them.' "

There were also many bad characters in the parish,
of debased habits, who chose for themselves, in the
way of recreation, only " the lowest and coarsest forms
of self-indulgence," and whom he thought it would be
very desirable to endeavour to reclaim " by putting
within their reach higher pleasures, and getting them
to take an interest in gratifying and agreeable ex-
hibitions."

It was this feeling which led to the display of fireworks, and other amusements which he provided for
them soon after his arrival, as before spoken of; and,
after a longer residence among his people, the same
feeling led further to the idea of making excursions
to places of interest within moderate distance, as,
" when properly looked after, and subjected to due
control, among some of the very best means at our
command for affording rational and agreeable recreation."

Moreover, recreations of this kind instruct as well
as amuse. We know how narrow and confined are the
ideas of a poor man, who has seldom been beyond
the precincts of his own village. He can hardly form
to himself any correct notion of other places, some of
which he has perhaps heard of, but none of which he
has visited. Churches and other buildings—woods,
hills, and rivers, and other natural productions,—are
all measured by what he sees in his own neighbourhood, and dwarfed down in his imagination to their
standard. And not only this, but he is equally igno-

rant of the ways and habits of other places. He becomes prejudiced in favour of customs to which he has been always used, and is slow to adopt any improvements suggested to him by those better educated than himself. Professor Henslow was desirous of rubbing off the rust that had been growing for years over the minds of the Hitcham people from this cause, and opening to them not merely agreeable sights, but sources of knowledge from which they had been previously quite shut out.

The first excursion was to Ipswich in 1848. A party of about forty were invited to accompany the Professor, each having "also permission to nominate a companion. Two or three of the farmers asked to be allowed to join." The following is Professor Henslow's own account of this excursion, and the one made the following year.

"They visited the Museum (at Ipswich), and inspected the extensive iron-works of the Messrs. Ransome; and all returned well satisfied with the treatment they had experienced. The orderly behaviour of the entire party was so perfectly satisfactory during the whole of this somewhat experimental trip, that I determined the following year to extend our operations. All who had been of the first party were now invited, each with permission to nominate a companion. I retained a *veto* in issuing the tickets, but found no occasion to exercise it in any case where application was made to me according to the prescribed rule. We were joined on this occasion by several of the farmers, to whom the good conduct of the party on the previous year had been reported. Our party now had reached to

170, including members of my own family, and all my
servants.   The proposal was to visit Harwich, and to
take a peep at the sea.   The party were not prepared
for all they had to expect.   Scarcely any of them had
ever seen the sea, or been in a boat.   Dread of the
water was a predominant feeling, and some of the more
faint-hearted stayed at home from this cause, though
longing to join us.   Even the railway was a novelty
to several, and there was much whistling and shout-
ing provoked on immerging into the darkness of the
tunnel."

The two excursions above spoken of, were undertaken mainly for the sake and advantage of a benefit
club, the members of which defrayed the expense by
devoting to that purpose the money that would have
been otherwise spent in their anniversary club dinner.
The third excursion, and the succeeding ones, were of
a more general character, the expenses being defrayed
chiefly by Professor Henslow's setting aside in like
manner the money which would have otherwise gone
towards his tithe-audit dinner.   The circumstances
under which this arrangement was made, deserve to be
stated more particularly, as affording another proof of
his moral courage, as well of his good sense.

It had been the custom at Hitcham, when he came,
as in some other parishes in Suffolk, for the rector to
give an annual tithe-dinner to the farmers in one of
the public-houses in the village.   These dinners appear
to have often led to much intemperance, and to have
been of a character not very creditable, either to the
host or to the parties whom he invited.   Professor
Henslow thought it very desirable to put an end to

them, and to substitute in their place something that might be productive of good instead of evil. Accordingly, in an address to the parish in September, 1849, he announced his intention of discontinuing the dinner, and applying the money to the excursions in question. The sum thus saved was not, indeed, sufficient for defraying all the expenses of these trips, but it was augmented by other small sums willingly subscribed by the people themselves, and by the rector's own friends, and the whole went towards what was from that time called the "Recreation Fund." The change thus introduced was probably not much approved of by some who had been in the habit of indulging at the late rector's table. But this was not a point which Professor Henslow stopped to consider. He at once determined what, in such a state of things, was the right and most judicious course to pursue; and if he incurred any ill-will in the first carrying out his plans, it gave way in the end to a unanimous feeling of satisfaction and gratitude throughout the parish, when the advantages resulting from the scheme he had set on foot were more duly appreciated.

It was an important element in these excursions, that the poor should contribute something towards the expense, which they seem to have done without the least grudging, deeply sensible of the advantage to themselves of the day's amusement. It shows the high estimation in which they were held by the labouring classes, when they were willing to pay in some instances eighteen-pence per head, independently of the loss of a day's work.

And Professor Henslow's good-nature, his desire to provide innocent and rational recreation for his parishioners, his delight in seeing them made happy for the day, was never more conspicuous than on these occasions. He spared no pains to make the excursions successful as well as entertaining. The whole plan was well considered beforehand, and every part of it systematically laid down, so that there should be no delays, no mistakes, no confusion. He sometimes even went so far as to " print and circulate plans of the route, with illustrations that should serve for reminiscences of the chief objects worth seeing; he arranged with the railway directors for cheap trains, and with public and private individuals for admission to interesting places; and most generously were his exertions everywhere seconded by all parties."

The first excursion, after the establishment of the "Recreation Fund," was to Ipswich and Landguard Fort in 1850; and the following copy of the Regulations issued by handbill on the occasion, is worth preserving, as showing the good-humoured spirit in which Professor Henslow got the excursions up.

HITCHAM EXCURSION TO IPSWICH AND LANDGUARD
FORT, ON TUESDAY, 30TH JULY, 1850.

I. According to my expressed determination last year, I am making arrangements for a village excursion to Ipswich and Landguard Fort; having received assurances that we shall be welcomed at both those places.

II. The successful issue of such an excursion (provided no untoward accident should occur) will mainly depend upon a general attention to a few rules, which I here propose to those who wish to accompany me.

III. Every one is to be in good humour, accommodating towards all, and especially attentive to the ladies of the party. If the weather should prove unpropitious, every one is to make the best of it, and not to complain more than he can possibly help.

IV. Those who are invited on the present occasion are the occupiers of farms in Hitcham, the resident members of the Stoke and Melford Club, and those who attend the adult class on Saturdays. Every one who joins the party may also name a friend, residing in Hitcham, for whose good conduct he will be responsible.

V. The tickets to be issued will be limited to 200, at 1s. 6d. each. All applications to be made at the Rectory on or before Friday the 26th. Should any tickets remain unapplied for after that day by the parties invited, those who have already received tickets may apply for more for other friends. By this arrangement I hope to accommodate all, or nearly all, who may be wishing to take part in the excursion.

VI. The party are to assemble on the platform at Stowmarket by half-past eight.

VII. As the object of the party is not to be mere eating and drinking, but wholesome recreation to body and mind, the refreshments provided will consist of bread, cheese, butter, cake, with lemonade, and one or two pints of beer for those who may apply for an order to that effect when they receive their tickets. These orders for beer are not to be transferred to other parties; and if not needed are to be returned.

VIII. Should any one be prevented at the last moment from joining the party, 1s. of the money paid will be returned; and the remaining 6d. will be appropriated towards the expenses that will have been incurred, upon the supposition that such person would have been of the party.

IX. Every one must contrive for himself how to get from Hitcham to Stowmarket, and back again. If he is not able to persuade any one to give him a lift, he must consent (as others have done before) to wear out a little shoe-leather.

J. S. HENSLOW.

Other excursions in subsequent years were made to Norwich, Cambridge, and Felixstow. In 1851, the year of the Great Exhibition, he even took a small party to London, where he kept them three days, feasting their eyes not only with all the astonishing productions of art and science that were brought together in that palace of glass, itself no small object of attraction; but with a sight of the Zoological Gardens, the Polytechnic, the Gardens at Kew, and various other lions in the metropolis.

Few of the excursions appear to have been planned and carried out with so much precision and regularity, as the one to Cambridge on 27th July, 1854. In this instance, a " Programme," or small pocket guide of eleven duodecimo pages, was actually drawn up and printed expressly for the occasion, in which all the colleges, museums, and other buildings and objects of interest in the town and University, were marked down in the order in which they were to be visited, followed by short notices of the more remarkable ones, and a general description of the ways and practices of college life; the dates of erection of the several colleges, with their respective numbers of resident members, besides various other particulars too numerous to be stated here.* To give some idea of the pains taken, and the cost incurred, by Professor Henslow, to instruct and amuse the whole party, it may be added that there were distributed among them for the occasion :—

250 of the above programmes.

36 Maps of Cambridge.

50 sets of plates of twelve colleges.

More than 100 single plates of ditto.

200 plates of some of the more remarkable plants in the Botanic Garden.

The above excursions were carried on for eight years in succession, when it was found necessary to stop them, in consequence of some new regulations on the part of the railway companies, which came in the way of the required accommodation. They had, however, been productive of the highest advantages; and it may be imagined with what delight and astonishment unlettered and untravelled villagers must have gazed upon the scenes and wonders thus laid open to their inspection. Nor could they fail to pick up much information under the leadership of the Professor, who

* A copy of the above little Tract is preserved in the Public Library at Cambridge.

would at each place stop from time to time, and gather
around him a flock of eager listeners, give his "lecturets," and explain to them everything they saw.
That the impressions left upon their hearts and minds
by those days, served in some measure to wean them
from the debasing pleasures in which alone they had
found any gratification when the new rector first came
among them, may be reasonably inferred from the
remark which a labourer once made to Professor
Henslow:—" Our heads would not be so full of drink,"
he said, " if we had such things as these to occupy our
minds." He was alluding to the objects in a museum
which had been visited in one of the village excursions.
Similar excursions have indeed been occasionally got
up by clergymen in other parishes for schools, but
seldom perhaps for adults on the same scale, or to as
great distances. Nearly 200 of the inhabitants of
Hitcham, men and women, are said to have sometimes
joined on these occasions; yet, notwithstanding these
numbers, whom Professor Henslow had to marshal
and keep in order as well as entertain, they were so
well disciplined, that there was never any accident,
nor any act of impropriety committed, to mar the enjoyment of the day, or to pain the feelings of the kind
and indulgent friend who had procured it for them.
It was " after one of these trips (that to Cambridge)
that the farmers of the parish, unable to withhold any
longer some expression of admiration, united in pre-
senting the Professor with a silver cup."

Such are some of the excellent and useful schemes
which Professor Henslow set on foot for improving the
social and moral status of his parishioners. They had

occupied much of his attention from the time of his
first entering the parish, and it was only by continued
perseverance, combined with a rare degree of moral
courage and patient bearing towards those who opposed
him, that he at length carried them out with that
entire success they met with in the end.

CHAPTER V.

MINISTERIAL DUTIES AND MINISTERIAL CHARACTER.

It is unnecessary to speak of other institutions Professor Henslow originated in the village, such as the Wife's Society, Coal and Clothing Clubs, Medical Club, Loan and Blanket Fund, &c., none of which existed in the place when he first came to it. These are now so common in all well-regulated parishes, that they call for no particular remark here.

But it would be a great omission not to speak of him in his strictly ministerial character, and to say something of the way in which he discharged the spiritual duties that attached to him as a parish priest. And it becomes the more necessary to do this, from its having been sometimes insinuated, if not openly alleged, that, while zealous in his endeavours to improve the social position of his people, and to impart to them secular knowledge, he neglected their religious teaching. It was hinted that he might have reclaimed some from criminal pursuits, and removed the ignorance which stood much in the way of their moral advancement; he might have devised schemes for winning them generally over to more steady and industrious habits, but that all this was not the "one thing needful;" and that the "one thing needful," if not entirely kept back, was at least not brought sufficiently forward. In short, the Parish Priest was thought to be too much merged in the Professor.

It is not surprising, perhaps, that such opinions, though very unfairly entertained, should have got abroad; nor is it difficult in some measure to account for them. The public generally heard of Professor Henslow simply *as* a Professor, who not only went about lecturing in the different towns of Suffolk, but who had introduced science into the village school. Botany was taught there, and no other instance had been ever heard of in which botany was taught to village children. Naturally, therefore, it became a matter of notoriety in the neighbourhood. It gained the public attention. It was thought that the clergyman who did this must attach extraordinary importance to such a subordinate acquirement, from which there was an easy step to the conclusion that higher subjects must be left out, or insufficiently taught, to make way for it. That this, however, was not the case, that the religious instruction given in the Hitcham school was quite equal to any imparted to poor children in village schools in general, might be plainly seen by those who attended the examinations. In addition to the weekly examinations in the school-room, there was a yearly examination of the children on the lawn before the rectory-house, the religious part of which was occasionally conducted by some of the neighbouring clergy. At these times the children always showed a competent acquaintance with Scripture, and with those chief truths upon which the Christian life is based. There was likewise, of course, the Sunday schooling before the services, in which the lessons were entirely of a religious character, and given by the members of Professor Henslow's own family.

And if botany was introduced into the school, it was not for the sake of the accomplishment itself, but, as before stated, from its being a subject well calculated to educe the faculties of the children, and to sharpen their senses, with reference to the uses to which they would have to put them when they grew up. Professor Henslow himself says, in a sermon preached in 1855, in aid of the parish school, " whilst, on the one hand, I am no advocate for imparting, merely for the sake of *instruction*, any knowledge which is unlikely to benefit those to whom it is given, I would never consent to abandon any form of teaching which is adapted to *educate* and *improve* those mental faculties which God has bestowed upon children expressly that they may be rendered useful to their progress in after-life." Nor should it be forgotten that the acquirement of this branch of knowledge was entirely voluntary. In no instance were the lessons made compulsory. Moreover, the circumstance of the greater part of the botanical work being done by the children out of school hours, removes surely all further objection to this science being taught them. Proficients in it as some of them became, it was, after all, simply one hour and a half to two hours in each week that was given up to the subject in the school-room itself.

But that the above remarks on the Hitcham school
may not rest entirely on my own judgment, I am glad
to have it in my power to subjoin the opinions of those
better able than myself to speak of its real character
and merits. For several years previous to the death
of Professor Henslow, the school was annually visited
by a Government Inspector. These Inspectors were
not always the same. But I am told they all expressed
decided approbation of the introduction of Botany into
the school (the opinion of one on this point has been
already given), while they found no such deficiency in
the religious knowledge of the children, as might be
attributed to the insufficient teaching of Scripture.

The Assistant Inspector before spoken of makes the
following remarks to me on this subject :—" I had no
reason to think that the botanical lessons interfered
with a due study of the usual subjects of a National
School. Independently of the botany, the Hitcham
school ranked *well* among the better class of rural
schools in the district I inspected. The Professor gave
a just weight to *all* the subjects of the school, in my
opinion. . . . . I have reason to think that the proportion which the children's knowledge of Scripture
History, of the plain doctrines of Christianity, and of
the Church Catechism, bore to the secular subjects of
the school, was quite on a par with that of other like
schools in similarly-constituted parishes." He adds,
" It is a fact that Professor Henslow's National School
produced more candidates as pupil teachers than the
school required for itself, and so they furnished another
school, if not more schools than one."

The above is in a letter to myself. In his official

Report on Hitcham School to the Committee of Council on Education for 1858, I find the following statement by the same Inspector :—" A very nice village school of intelligent and cheerful children. Reading good, *i. e.* expressive as well as correct. Religious instruction good. Writing fair. Arithmetic pretty fair. Needlework fair. Extra subjects pretty fair, and among them Botany excellent, this last being most thoroughly yet simply taught, and by such a system that there can be no cram. As far as a child goes, it must *know* what it does. The good moral effect of this study on the minds of the children is very apparent."

In conclusion, on this subject, it may be stated that, on one occasion, an Inspector of Training Schools, having heard of the reputation of Hitcham School, though not called upon by office or duty to visit it, came down on purpose to see the system of teaching Botany there, and highly approved of what was done.

In reference to the general spiritual ministrations of Professor Henslow in his parish, it may be remarked that these were not the parts of his character and daily life which would come before the public at large any more than the religious teaching of his school children. They are what every conscientious clergyman attends to, and which therefore excite no particular notice beyond the immediate neighbourhood in which they are carried on. The private intercourse especially between a clergyman and his parishioners goes on noiselessly in the world. Its secret influence for good is felt only by those for whose benefit that influence is exerted.

That Professor Henslow was anxious for the growth of grace among his people, and for their observance of the exercises of religion upon a right principle, appeared clearly in many of the sermons he preached. That he himself neglected no means by which he might raise the tone of Christian feeling in the parish is shown by the result of his endeavours to this end. There are persons in the village of Hitcham of good Christian life, who attribute all their religious knowledge to his teaching. His people were steady in their adherence to him,—a steadiness which much increased during the later years of his ministry,—and which arose mainly from his own steadiness in attending to all their wants, spiritual as well as temporal. He was almost constantly resident among them. He was very seldom away, except during the period of his lectures at the University, and even at such times he often returned upon the Saturday in each week for the Sunday duty. For more than twelve years he was not absent from Hitcham for a single Sunday. He was unwilling to leave his congregation in the hands of strangers, whose teaching might not be in accordance with his own views, and which might have the effect of unsettling the minds of his parishioners. For the same reason he mostly refused his pulpit, when solicited, as he occasionally was, by other parties, to allow them to preach in his church for missionary and such purposes. He took as warm an interest in the mission-field itself as any man. He was as anxious for the extension of Christianity in foreign lands as in our own country. But he alleged that his people were extremely ignorant, that he was a better judge than

strangers as to the best way of drawing their attention
to that subject, and that, as often as he saw fit, he himself preached missionary and other charity sermons,
to which the poor always responded as liberally as
their means allowed.

There was latterly the more occasion for Professor
Henslow to keep a strict watch over the spiritual
interests of his flock from the spread of much fana-
ticism in the parish through revivalist meetings. These
meetings had become very frequent in the neighbourhood the year before he died, and had been attended
by the most distressing results. Whole families had
been pained by the influence exercised upon the minds
of some of their members by the pernicious doctrines
inculcated at these meetings, which the parties had
been tempted to join. In the case of certain individuals the excitement thereby induced reached to
such an extent as to end in insanity. Professor
Henslow did all he could to check the evil effects of
such mischievous teaching. He spoke strongly upon
the subject in the pulpit, besides giving his advice in
private to such as needed it. One young woman,
who had been much talked to by a Revivalist, and for
a short time made very wretched, expressed to one of
the members of his family how much she had had her
fears relieved by reflecting on what her minister had
told her, and which she found in exact accordance with
the Bible. Many others said as much. This shows
his readiness, as occasion offered, to " reprove, rebuke,
exhort, with all long-suffering and doctrine." Nor was
he less forward in administering consolation and support in all cases of distress and suffering, whether

mental or bodily. It was affirmed by some that, in times of trouble, when they had no one else to fly to, they never failed to find in him a sympathizing friend. One poor woman remarked to a sister of the Professor's, what a comfort and blessing his ministerial visits had been to a sick son of hers, who was dying of a lingering disease; as they had also been to three other children, whom she lost after they had grown up. Of one she said, "He saw her every day for a fortnight before she died." The Revivalists, who on one occasion held a prayer-meeting in this woman's house, during the illness of the son above alluded to, made much of the happy state of this young man's mind, as if it had been due to their teaching. Whereas she said, "her son could get no rest after it, and resisted all their efforts to have another." She added, "I read the same prayers to him Mr. Henslow did: we don't hold to praying without a book; we can't understand them so well. I am often out nursing, and I always read those prayers to the sick night and morning." This woman was almost the first of the Hitcham people known to Professor Henslow and his family,— a simple-minded, affectionate creature, and one who would tell the honest truth; her testimony, therefore, to the assiduity with which he watched over the sick members of his flock may be fully relied upon. Nor was it the sick only that he visited. He saw all his parishioners in their turns; he listened to every call for assistance and advice, and gave free access to those who sought him out for any occasional purpose at his own house.

Professor Henslow belonged to no particular party

in the Church: he was neither High nor Low in his
religious opinions, and religious teaching. He was
no controversialist; he adhered to no particular theo-
logical system, and had not of course, with his multifarious pursuits, read so deeply in divinity as many
other clergymen.   He rather seemed to take his stand
upon the Bible itself, which he daily studied, and with
which he was thoroughly acquainted. The groundwork of his ministerial teaching was in strict agreement with the Sacred Volume, as it was also with the
Articles of our Church.

At one period of his life he gave much of his attention to the study of the prophetic books of the Sacred
Scriptures, especially the book of Revelation; and he
joined a small circle of friends in the University, who
met together occasionally to discuss some of the more
obscure passages in those books.   It was whilst engaged in those studies that he preached in Great St.
Mary's Church, Cambridge, before the University, a
sermon " On the first and second Resurrection," which
caused much sensation, from the circumstance of his
being to such a degree excited by the subject, that he
burst into tears in the middle of his discourse, and for
some time was unable to proceed. This sermon he
afterwards published.   Later in life, however, he in a
great measure abandoned these inquiries, being satisfied
that such mysteries were not to be explained without
further light and knowledge than was to be attained
at present; that their right meaning could only be
evolved as the day drew on for their fulfilment; and
that it was far more profitable to confine himself
chiefly to those parts of the Bible which had a plain

practical bearing upon every-day life, which yielded
the surest supplies of strength and consolation to the
Christian, and which also furnished the Christian
minister with the most important topics for public
preaching and private exhortation.

His Hitcham sermons were all more or less practical. They were composed rapidly, and written off-
hand. Yet there was nothing in them to give the
idea of their having been written in over-haste, or
without sufficient thought. They were alike sound in
doctrine and reasoning ; the language clear and well-
chosen. There was a substantive weight about them,
arising from the philosophical turn of his mind, far
beyond what appears in a large number of published
sermons, upon which probably much more labour has
been bestowed. There were no vapid, meaningless sen-
tences ;—no common-places, no repetitions ;—no eking
out of the matter to bring the sermon to its proper
length. Scarcely a word could be struck out, without
either impairing the sense or weakening the argument.
They were destitute of both metaphor and ornament.
On the whole the style and language nearly assimilated to the style and language of the late Arch-
deacon Hare, though free from some of the peculiarities of that author. If deficient in anything, Professor
Henslow's sermons were deficient in warmth and ani-
mation as regards the way in which he treated his
subject. Truths, doctrinally important, and highly
necessary to be enforced, were often too abstractedly
considered ; little being said to lead his hearers to
make a personal application to themselves of what
they heard. Perhaps his sermons would have been

more attractive if more had been yielded to the circumstances and condition of his hearers. They were
addressed to a congregation for the most part but
scantily educated, and yet they required in some
measure an educated mind,—not so much to compre-
hend them,—as to be impressed by the lessons they
conveyed. They were directed more to the understanding than to the heart. The common people, it is
well known, like to have their feelings excited by the
preacher, and unless this is done to a certain extent,
though they may listen attentively to the sermon
while being delivered, they seldom reflect much on it
afterwards.

For this reason, it may be doubted whether his
public ministrations in the church had all the effect
upon the poorer part of his flock which his private
ministrations had, and the truly Christian example he
displayed in his every-day life. In like manner, his
preaching was not after the best models of pulpit
eloquence. His voice was loud and sonorous — at
times over-loud—with strong emphasis upon particular
words and passages, where there did not seem to be
occasion for the stress he laid upon them. Preaching,
certainly, was not natural to him in the way that
lecturing was, in which he excelled. His manner of
reading the prayers and lessons was much like that of
his preaching. Yet, with all this, there was an earnestness about him which showed that his own heart
was thoroughly engaged in the work, and that he
exerted himself to the full extent of his powers.
Whatever there might be in his mode of doing the
duty capable of improvement, there was at least no

coldness or formalism in it; no ground for the charge
that the spiritual interests of his people did not hold
the first place in his consideration for their welfare.

It should be remarked, too, that what has been said
above respecting the matter of his sermons, applied
chiefly to sermons on general subjects. For he was
much in the habit of that excellent practice of what
has been called "local preaching," and which Paley
has made the subject of a particular charge, strongly
recommending it to the clergy. That is to say, he
would adapt his sermons to the prevailing tone of
thought and opinion among his parishioners, and the
particular circumstances of the place, according as
they arose. He would not fail to notice anything
that called for notice, and make it an occasion of
advice, warning, censure, or encouragement, as the
case might be. If any erroneous impressions or opinions got abroad, in which the interests of religion
were concerned, as in the instance of revivalism before
alluded to, he would correct them publicly. He would
sometimes plead in behalf of the labouring poor—
sometimes take up the cause of their employers. He
would, of course, strongly advocate the importance of
educating poor children, and insist upon the claims the
school had upon the support of those who were able to
assist it with their money, as also upon the duty of
parents to remove all hindrances in the way of their
children attending it regularly. On this latter point,
he observes that,—

" The greatest obstacle with which they have to
contend, who are desirous of seeing the youth of all
classes duly educated, lies in the ignorance or indiffe-

rence of many parents to what education really means.
Wholly uneducated persons are necessarily rendered incapable of duly appreciating the blessings which a good
education is calculated to confer upon every one who
does not abandon himself to vicious practices, and so
become both the willing slave of Satan in this world,
and the foredoomed citizen of Hell hereafter. They too
often fancy themselves not bound by any duty to their
children to make the necessary sacrifices (however
trifling these may be) to send them to school, or to
keep them steadily at their books for sufficient time to
profit by what they have to learn. Hence the most
trifling excuses are made for interrupting their studies.
They should understand that every such interruption
weakens the efforts essential to maintain a steady
advance."

He seems to have been favourable towards a compulsory attendance of village children at school,
between certain ages and at stated times, and much
lamented the farmers taking them away so soon to
work; concluding the sermon, from which the above
passage is quoted, thus :—

"Although I trust something will soon be done, on
the part of Government, to extend those advantages to
the children of agricultural labourers, which the children employed in factories already enjoy; and although
I fear comparatively little good can be effected until
some general scheme of national education shall be
devised, still there can be no reason, but the contrary,
for our not attempting to educate as many children as
may be allowed, by parents or masters, to attend our
village schools, even though it be for too short a

period for many of them to receive as much instruction as they ought."

In the earlier part of his ministerial career at Hitcham, there was much unreasonable opposition on the part of the farmers, as already spoken of, to some of the schemes he devised for the good and improvement of the parish. This, too, was often met, as it deserved, with a just rebuke from the pulpit, conveyed in temperate language, but, at the same time, strongly expressive of his surprise and grief to witness the attempts made to hinder him in the discharge of his duty, on the part of those professing to be under the influence of Christian feelings and Christian principles. If the censure did not produce any immediate change in the outward behaviour of his opponents, it must at least have had the effect, in some instances, of making them inwardly ashamed of the course they were pursuing.

Or, again, if any unusual occurrence had taken place in the neighbourhood—a sudden or a violent death, or an accident to life or property, or if any crime had been committed, or his people had become addicted to any moral irregularity—such things were sure not to be passed over in his discourses. In like manner, in seasons of particular distress and hardship, he would enforce the moral right which the sufferers had to relief from their richer neighbours; at the same time reminding his hearers of an over-ruling Providence, whose care extended to rich and poor alike, and who ordered all events for the good of those who look up to and trust in God.

No sermons are more attended to than those of the

above character, for they relate to matters which all
understand, and in which all are interested; and if
drawn up in Christian language, and made the occasion of enforcing Christian truths and precepts, none
are more profitable to the hearers.   Moreover, as they
can only be preached by clergymen who are habitually
resident among their flock, and who are thoroughly
acquainted with all that goes on in the parish, they
testify to the conscientious solicitude felt for the moral
and religious advancement of the people committed to
their charge.

The sermons of this kind, which Professor Henslow
preached, are excellent in their way, and written in the
plainest and most familiar style.   Nor was he content
merely with preaching them.   He would often have
them printed for general circulation in the parish, so
that they might not only be read and thought of afterwards at home, as well as listened to in church, but be
the means of conveying the lessons they contained to
those who were not present when they were delivered.

There is one sermon, especially, on "Sudden Death,"
preached on the occasion of a man having died, as it
would seem, by exposure to cold during a fit of drunkenness, which for its plain language, sound practical
advice, and solemn warnings, must have been heard
with deep attention.   The same may be said of another
sermon preached " on the occasion of the Irish
Famine," some passages of which contain as forcible
an appeal to the feelings, as can possibly be desired.
These two sermons, as well as others that might be
mentioned, show the extent to which the preacher's
own heart must have been warmed with the subject

before giving utterance to the sentiments he spoke. They are such as no clergyman could have penned, who did not feel to the full the responsibility of his sacred office — who was not truly alive to the importance of rescuing souls from the ways of sin, and bringing them back to God.

After all that has been said in reference to Professor Henslow's public ministrations in the church and parish, it might have been thought unnecessary to touch upon the hold and influence which religion had upon his own heart. Yet here, too, doubts have been expressed, and insinuations thrown out, disparaging to his character as a Christian, independently of his profession as a clergyman. Not that any have denied, for a moment, the unblemished excellence of his moral life, or the earnestness with which he laboured to do good in his generation, and to make himself useful in the world. But it was surmised by some that this was mainly the result of temperament, and his peculiar turn of mind in the direction in which his chief occupations lay. He was thought to be too fond of science to busy himself with, or, indeed, to find much time for, personal religion. The philosophic cast of his thoughts, as judged of by his conversation, seemed to forbid ordinary association with the subject. He was strong in himself; and it led to the idea that he felt no need of continual recourse to those private devotional exercises, which are the stay and support of all true Christians.

Nothing can be further from the truth than all this. Those who knew him best, know him to have been one of the most thoroughly spiritually-minded men of his

profession.   Nor is it possible for the public generally
to be competent judges of the extent to which a man
cultivates personal religion, which is so much a thing
of the heart, when there is nothing at variance with it
in his outward behaviour.   Professor Henslow was not
a man who talked about religion in mixed society, or in
ordinary conversation.   He was no hypocrite ; and he
shrank from assimilating himself in the smallest degree
with those who have the name of the Saviour for ever
upon their lips, without any corresponding sanctity in
their lives.   He was content to let his religion appear
in the faithful discharge of the duties of his calling,
and in that universal good-will and charity which he
bore to others.   And although he might not usually
converse on religious topics, he could and did do so when
a fit occasion offered.   More often it was his friends
rather than himself that led the conversation another
way.   His attainments in science were so varied and
extensive, his knowledge so accurate, and his reputation as an authority in all scientific matters so widely
established, that persons were glad to refer to him,
and to get instruction from him whenever they could.
He was equally ready to impart the information they
sought.   He imparted it, too, in so clear and simple a
manner, and with such entire freedom from all affecta-
tion and pride, that no one felt afraid to approach him
for the purpose of getting either his opinion or advice.
Hence it happened that, wherever he was, especially
when there were present strangers who knew him by
name but who had never been in his society before,
questions were constantly being put to him connected
with the pursuits which he followed up with so

much zeal and success. But if religion instead of
science were made the subject of conversation, as it
sometimes was among his intimate acquaintance, he
was quite as ready as others to join here also. Mr.
Darwin, in his communication to this memoir before
given, states that Professor Henslow had talked to him
" on all subjects, including his deep sense of religion,
and was entirely open." Mr. Berkeley also informs me
that he had " had many long conversations with him
upon his religious views."   I have also had the same
myself; and on occasions when any particular passages of Scripture were brought under his consideration, he would not only show himself fully competent
to discuss their right interpretation, but he would
afford convincing proof to his hearers, that he was as
familiar with the Book of Revelation as he was with
the Book of Nature.

The real truth is, that he had not only thought
deeply upon religion, but had habitually sought its
help and guidance, and especially on all the trying
occasions of life. Many instances are known to his
own family of the secret communion he kept up with
his Maker. His devotional ardour was, indeed, plainly
manifested during the severe illness which preceded
his death. But long before this, when he was yet in
full health and vigour, there were circumstances at
times which elicited from him those feelings which can
alone exist in the breast of a truly religious man. He
himself referred to a time some years ago, when he
" prayed God, that if he could be of any use in advancing His scheme of salvation, He would make use
of him for that purpose," adding, " perhaps He has."

And who that reflects on the success that attended his various undertakings for the good of his parish, can doubt that it was mainly due to his earnest prayers for spiritual direction? He spoke, too, of a special occasion for prayer shortly previous to his getting the crown living of Hitcham. It had been under consideration whether he should not be appointed to the see of Norwich, the bishoprick of which was then vacant, instead of to any lower preferment in the church. On hearing this, of which he had certain information from a friend, he retired into his chamber, and fervently on his knees prayed for some time that he might never be called to any such high office, for the duties of which he felt himself quite unfit, and that he might not be tempted to accept it if offered to him. When he found afterwards that he was to have the living of Hitcham, and not the bishoprick, he thanked God for the issue, and regarded it as in answer to his prayers.

Allusion has been already made to the period when his mind was in a highly excited state from being too exclusively directed to the subject of the prophecies. After having become convinced of the unprofitableness of such studies, when carried to excess, he seems to have settled down to calmer feelings upon religion, and to have restrained himself from giving utterance in public to anything that might savour of enthusiasm, or which might seem to indicate an overheated imagination. But though he saw the mistakes he then fell into, he was only cautious afterwards not to fall into the like again. The strong impressions made upon him at that time were never so effaced as

to beget the opposite extreme of lukewarmness. Scripture was still studied, only for more practical purposes ;
the flame of devotion was kept up, only it burnt more
steadily; and this sobered condition of mind was preserved for the remainder of his life. Some idea of
the degree to which his feelings were worked up at
the above period may be formed from the way in which
he himself spoke of it during his last illness. The
conversation having turned to the subject of the fana-
ticism which was then gaining ground in the parish
from the influence of the revivalists, he referred to
the day " when his own imagination had been misled,
and when the conflict of mind was so dreadful that he
believed he was one night nearly insane ; " adding that
" he prayed fervently, searched Scripture for himself,
and asked advice from friends qualified to give it :
among others, he went to consult Irving, whose
avowed surprise at being unable to raise, by his
prayers, a dead person, whose coffin he showed him,
brought him to his senses." One who retained so
lively a recollection of the intensity with which religion had once exerted its influence upon him, was
not likely, at any subsequent period of life, to regard
religion as a vain thing, or that might be lightly dealt
with, however much he might regret his former errors,
or however seldom he might openly reveal what his
altered sentiments upon the subject were.

He seemed always to take a high view of the
Christian Dispensation, regarding it as a " law of
liberty " to a greater extent than many are wont to
allow. He often spoke of the privileges granted to
Christians, which many so far from adopting would

bind themselves in the very bonds from which Christ had set them free. Thus fasting, when practised according to the letter, he regarded as a falling back to Judaism. The Christian's fast, in his estimation, was a fast from sin. " Meat commendeth us not to God." In like manner he considered the over-strictness with which some observe Sunday as showing a tendency to confound it with the Jewish Sabbath. He thought it foolish to lay restrictions upon things in themselves innocent, and that were not made the occasion of sin. " Unto the pure all things are pure." This may partly account for the opinions of some that he was not earnest in religion. He was earnest in respect of all that was inward and spiritual; but outward observances, when not positively enjoined, and which seemed rather to belong to the Old Dispensation, he felt himself at liberty to pass by. " Where the spirit of the Lord is, there is liberty."

One great feature in his religion which always struck me, was the strength of his faith. There was with him a complete realization of those great truths of the Christian Revelation, upon which all right practice is based, and which supply the highest motives to action with the true believer. He was so fully assured of the certainty with which all the promised fruits of a sincere profession would follow, in the case of those whose Christianity was real and in the heart, that, from first to last, he entered upon the discharge of life's duties, and looked forward to life's end, without the slightest misgivings. " Is it possible," he asks, " that any sinner (using the term as applicable to all the sons of Adam) can be so privileged as to be able

to attain a state of mind so blessed, that he has no
longer any fear about what shall be his fate hereafter ?
This is not only possible to a chosen few, but it is the
high privilege of every one called to be a Christian." *
His was, indeed, the faith which seemed able to remove
mountains. It was his strong faith which encouraged
him to undertake what he did, much of which it might
have been thought by some hopeless to attempt ; and
which carried him through difficulties, at the sight of
which many would have turned back. It was the
same strong faith to which may be traced the equa-
nimity with which he took all the events and accidents
of life, and the calmness with which he reasoned with
opponents, whose hearts had not been brought under
the same principles as his own. And this faith, which
was so conspicuous in himself, he impressed on others.
He considered, with Dr. Arnold,† that " unbelief is at
the bottom of all our evil;" and the ground of his
public preaching, as well as of his private exhortations,
was—that if men would only believe, as they ought to
do upon the evidence they had received, they would
find no difficulty, by the help of Divine Grace, in doing
what was right, and they would never miss of their
reward. He says, in one of his manuscript sermons,
" If men would only believe, as readily as they are
inclined to hope this true," (that Christ has made a
sufficient sacrifice for the sins of the whole world,)
" sin would soon cease to have that strong hold by
which it retains so many in slavery. . . . . Unbelief
is the curse that weighs them down to hell. Men

* Extract from a MS. Sermon.
† Life and Correspondence. 8th Edit., vol. i., p. 311.

H

won't believe themselves to have been set free from sin. See how they persist in committing wilful sins." Again, " What we ought to be is set before us in Scripture, and that we may most certainly become (by God's help) if we will only believe it. But what we ought to be we never shall become if we will not accept the means and employ the methods by which God has decreed that we may attain success."

In equally strong language, on another occasion, does he dwell on faith as enabling us to bear the trials of life. Speaking of those who murmur and complain at God's afflictive dispensations, and who are ready to accuse Him of injustice, he describes such impiety " as a consequence of that sin which so easily besets us, a want of faith in God's promises. We do not believe that they are always fulfilled to us, in spite of whatever may be the seeming contradiction to their fulfilment. God cannot lie. His promises are fulfilled to believers under all circumstances whatever, even under the severest trials to which they can be subjected."

Yet with this strong faith there was a child-like humility of mind, which at once indicated whose disciple he was, and from what sources his strength had been derived. He did not talk of his doings, nor refer to the good he had been instrumental in effecting, as entitling him to any commendation. Of course his more open schemes of usefulness in his parish were known to all; but what he did in his private ministrations was known to few, and this was one reason why he was supposed to do so little in this way. In one instance, in which he was informed of a neighbour who had stated publicly that he, the neighbour, had been the means of converting a young man at Hitcham by means of revivalism, Professor Henslow, who knew that his conversion was due mainly to his own previous instruction given on the occasion of his visiting him during an affliction, very quietly observed, " —— has no right to take the credit of that," without any reference to what he had done himself.   His humility, too, was as much shown in matters of science as in matters of religion.   When he first came to Hitcham, an acquaintance of his in another part of the country happened to be in the neighbourhood at the time, and it was remarked to the latter by one of the residents there, that " they had been almost dreading the arrival of the learned Professor," thinking he would be un-approachable, from the high position he occupied in the scientific world.   " Oh," said the gentleman, " you need be under no fear, for I never knew a more humble-minded man in my life."   And it was not long after his arrival that they found how groundless their fears were. However learned he was, his learning was no bar to his mixing freely with all classes, while he condescended to instruct the most ignorant and illiterate who sought him out.   On all occasions he under-rated his own abilities.   He would listen to no praise or flattery whatever, and in his last illness gave particular directions that all letters that contained any should be burnt after his death.

Who will say that a man of such feelings and habits of mind as have been described in the preceding pages was not thoroughly imbued with the true spirit of Christianity ? that he did not live in the constant recollection of the Divine Presence, looking up to God

as the Disposer of events, desiring to be led by his
Providence and strengthened by his Grace, and seeking
alone to be accepted in the great day, when all human
actions shall be inquired into, and the inmost recesses
of the heart laid open, and " every man have praise of
God ? "

CHAPTER VI.

From Professor Henslow's long experience and great
popularity as a lecturer, he was often called upon to
lecture in some of the towns of Suffolk. Bildeston,
Hadleigh, Sudbury, Ipswich, and Bury St. Edmunds
had all the privilege, from time to time, of hearing his
addresses on different branches of science. He was
also occasionally asked to lecture in more distant
places, but, with the exception of his lectures at the
University, he generally declined going out of his
own county for this purpose. In London he lectured
once at Buckingham Palace, having been invited by
H.R.H. the Prince Consort to give a short course of
lectures on botany to the junior branches of the Royal
Family. "These lectures were delivered *vivâ voce;*
they were, in all respects, identical with those he was
in the habit of giving to his little Hitcham scholars;
and the same simple language and engaging demeanour
that had proved irresistible in the village, won over his
Royal audience to fixed attention and eager desire for
instruction."

No town, after Cambridge, is so much indebted to

Professor Henslow, as Ipswich. He will be long remembered there, not only as a lecturer, but for the services rendered by him to the Museum, which, from the time of his being elected President, was placed entirely under his direction, and which, for the admirable manner in which it is arranged by him, has justly acquired great celebrity.

This Museum was established in connection with a proposed system " for giving Instruction to the Working Classes in Ipswich in various branches of Science, and more especially in Natural History." At the first meeting of the subscribers, Professor Henslow had been elected an honorary member, and in March, 1848, he was requested to deliver an Introductory Lecture in the Museum, the first President being the Rev. W. Kirby, the celebrated entomologist.

This lecture is an excellent specimen of one of the many popular addresses which he delivered at different times at such Institutions. In it he draws the attention of his hearers to the advantages and great importance of science in general, as well as to the mutual bearing of the different branches of science on each other. He regrets how much it has been undervalued, not only by the illiterate and ignorant, but, in some cases, by those who have studied deeply other departments of human learning. Whereas, " scientific information has its bearings even on the occupations of the profound critic and learned divine." He observes that it sometimes helps us to a better understanding of Scripture itself, in certain passages of which our translators have made great mistakes, from their ignorance of the plants and animals found in those countries to which Scripture refers. He alludes to "those gross blunders which are so constantly made, even by otherwise well-educated persons, concerning the nature and origin of many of the natural objects with which we are daily familiar;" the "many popular errors about insects;" the prejudices entertained by our forefathers, and which still retain their hold upon so many minds, which a very moderate acquaintance with science would serve to remove; the frauds and impositions often palmed upon the poor, and, sometimes, even on "educated men," by quacks and other "ignorant pretenders." All these evils would be much lessened, if they did not quite disappear, were but natural history and the other allied sciences more duly appreciated and taught in our schools, and were there an "establishment of an efficient system of national education," which he so much longed to see. Yet, in this Address, as on many other occasions, he again and again reminds his hearers of the necessity of their making a certain "degree of mental exertion for mastering the first rudiments of science." He never encouraged any to believe that the mere listening to an amusing lecture, or reading popular books, would alone make a man a naturalist. These things do but "smooth the road to further inquiry." He must fixedly give his mind to the subject in other ways. He must observe diligently for himself, while his senses and faculties must both be duly disciplined to enable him to observe correctly, and to turn to a right account the knowledge he acquires. This is excellent advice.

Excellent, also, are the remarks with which he concludes this Address respecting the supposed interference of science with religion.

"Some persons," he says, "consider that these pursuits are calculated to exert a pernicious influence upon our hearts and minds, and that they have a tendency to lead us astray from higher and more worthy objects. Now, I would at once concede the point, to such as choose to moot it, that the highest estimate of the Divine Power and Godhead which may be obtained from our closest observation of the works of the creation, cannot procure for us the least and lowest of those spiritual graces which require (as we believe) an implicit confidence in God's Word. But after having granted thus much to our unscientific opponents (and they are avowedly our opponents), I would grant them no more. If they tell me (and I have been so told) that science is necessarily a snare and an obstruction to our spiritual progress, I tell them, in return, that I utterly deny and disbelieve their assertions. Whilst I freely admit some of them to be very far better judges than myself of how great a variety of accurate interpretations may be put upon the mere letter of God's Word, I claim full freedom for the exercise of private judgment concerning those spiritual things which the Bible is intended to teach me. This privilege I conceive to be the legitimate birthright of that weighty responsibility laid upon every man, to take heed to himself as to whether he is standing or falling to his own master. To all who would refuse us this high privilege (were they the most learned of men), we may safely reply in the

words of the patriarch, 'No doubt but ye are the
people, and wisdom shall die with you. But I have
understanding as well as you; I am not inferior to
you: yea, who knoweth not such things as these?'"
(Job xii. 2, 3.) "Can they, indeed, suppose it possible that a closer inspection of the lilies of the field
than they have an inclination to give them, will necessarily compel us to acknowledge that these good creatures of God have now lost somewhat in the comparison
of their clothing with that of the richest garments
ever woven for an earthly potentate? Do they,
indeed, fancy that our corrected estimate of God's
wisdom, and our improved appreciation of God's
power, obtained by a stricter investigation of His
works, must necessarily be tempting us to declare His
goodness to be no longer unsearchable, His ways no
longer past our finding out? What if they can hold
up to us the names of some men, glorious in the eyes
of the world on account of their scientific attainments,
who have, nevertheless, been disbelievers in God's
Word? how can they tell to what extent such unworthy infidelity had been fostered, perhaps engen-
dered, by the very oppositions, persecutions, and
obloquy to which those men had been subjected by a
prejudiced intolerance? May they not have been
seduced into infidelity by the very conduct of their
bigoted opponents, who would thus have thrust upon
them their own poor puerile notions of what science
ought to teach, but which their own more enlightened
understandings had shown them to be no otherwise
than gross absurdity? I do not say this in justification, or even in extenuation, of those infidel philoso-

phers who have rejected the Bible merely because they
could not accept the precise interpretations which their
opponents had chosen to put upon its teachings.
They might have done better for themselves. They
are responsible for not having tried to do so. But it
is a lawful question for us to ask, whether the too im-
patient incredulity of one party is at all worse than
the too uninquiring credulity of the other. Let us
be mutually tolerant, mutually confiding, and then, in
due course, we shall learn to see how impossible it is
that either the works of God, or the Word of God,
can ever be teaching us things contradictory to truth.
How far the patient study of all God's various methods
of revealing to us His holy will, may be actually necessary for mutually throwing light upon each other, I
will not venture to suggest. Of this I am quite sure,
that where the study of God's works is combined with
a sure faith in His Word, the former can in no respect
impair our spiritual possession of the life that now is,
or deprive us of the enjoyment of one jot or tittle of
those glorious promises which have assured to us a
blessed immortality."

The above Address, which was never published, was
delivered on the 9th of March, 1848. In July, 1850,
the death of the venerable Kirby left the Presidentship
of the Museum vacant, and Professor Henslow was installed as his successor in the chair at the ensuing
anniversary. On this occasion, the late Professor E.
Forbes came down to Ipswich, and gave a gratis lecture for the benefit of the Institution.* The new Pre-

* This circumstance is alluded to in the Life of Forbes,
lately published, as follows :—In Dec. 1850, " Edward Forbes

sident now exclusively superintended the arrangement of the collections. He gave up to them a great deal of his time and attention, not merely assisting with his advice, but taking his full share in the manual work that was called for, in order to turn the establishment to the best account, and to bring it to the highest state of usefulness.

Professor Henslow had thought very much on the subject of museums, especially such provincial museums of natural history as are now to be found in most of our large country towns. Having, too, himself collected largely in almost every department, and given up much of his consideration and spare time to the best mode of arranging and exhibiting objects for public inspection and public instruction, he was admirably qualified for directing others in this matter. The Ipswich Museum, as planned and arranged by him, and those under him, may be fairly taken as a model of what a country museum should be. It is too common in such institutions to receive everything that can be got together by gift or purchase, without rule or method, until a heterogeneous assemblage of natural objects is amassed, which may serve for purposes of amusement, but serve very imperfectly for purposes of instruction. Or the collection may be rich, perhaps, in certain particular

went down to Ipswich to give a gratis lecture, for the benefit of the Museum there, on its anniversary meeting." A party accompanied him, consisting of " Van Voorst, R. Taylor, Bowerbank, Wallich, Mitchell, and Lankester." " Henslow had to be installed as their new President." Forbes says, " I stayed with Ransome, who gave a great dinner-party (on the occasion), when it was pleasant to see the bishop, four or five clergymen, and a bushel of naturalists, all dining at the table of a quaker chemist and druggist."—Life, pp. 481, 482.

departments, and these may be well arranged, whilst
in all other departments it is more or less deficient.
Professor Henslow saw this mistake, and set himself
to correct it.   What a beginner of natural history
wants in a museum, is a general view of the productions of nature, so grouped as to show their mutual
relations, in respect of structure and the several periods
of the earth's history to which they belong, before
selecting any one class for his own study.   To obtain
this end it is quite sufficient to have for display a
single representative of each of the larger groups,
choosing such objects as may be considered most typical of the class or family to which they respectively
belong.   This will form an instructive series of specimens, which may be got together at no greater expense
than most local institutions of this kind may be supposed capable of bearing.   The shelves, too, of the
museum will not be unnecessarily crowded with a
multitude of closely-allied genera and species, which
only large buildings can contain.   If the space is very
limited, and the funds of the establishment small, a
museum may be nearly or entirely limited to such a
series of typical objects, and yet be extremely interesting as well as instructive.   If there is more room, the
extra space may be devoted to a special collection of
all such natural productions as occur in the neighbourhood, or in Great Britain generally, to the exclusion of
other countries. *

* Edward Forbes had formed very similar views to those of
Henslow respecting Provincial Museums. He complained that
they were " too often ambitious attempts at general collections,
and, necessarily, failures. Too many of them are little better
than curiosity shops. In their best aspect they are, with very

The Museum at Ipswich should be seen in order to
judge of the admirable way in which Professor Henslow has there carried out these views. Beginning
with the elements, so far as these are capable of exhibition in a museum, the collection proceeds with the
most important and characteristic simple minerals,
models of crystals, rocks, fossils, and other geological
specimens, all chosen so as best to illustrate the order
of strata, and arranged according to age; from these
it passes on to the vegetable and animal kingdoms,
every group being represented in its turn as far as
practicable, and a model or drawing, or a tracing
from some published figure, being introduced as a
substitute in those cases in which no specimen of the
required typical species could be procured. For the
guidance of others, Professor Henslow drew up, at the
request of the Natural History Section of the British
Association, a Report "On Typical Series of Objects
in Natural History, adapted to Local Museums," in
which a list is given of such species and objects as

few exceptions, far more costly and far less useful than they
ought to be."
He would have "a Provincial Natural History Museum to
consist of three departments, for which three spacious rooms,
if properly constructed, would suffice."
"1. *A Local Collection*, illustrative of the Zoology, Botany,
and Geology of the county or district.
"2. *A Teaching Collection*, consisting of carefully-selected,
well-arranged, and thoroughly-labelled types of the Classes,
Orders, Families, and leading genera of Animals and Plants,
of the series of geological formations and their characteristic
fossils, and of Minerals—no superfluous specimens to be admitted.
"3. *A Miscellaneous Collection*, including fine or rare specimens of exotic productions not necessary for the teaching
collection, remarkable curiosities, and ethnological illustrations."—*Memoir of Edw. Forbes*, pp. 513, 514.

form the best selection for such a purpose, the same
having been supplied to him by different naturalists
who had made different departments of the subject
their particular study. This paper was published in
the Report of the Association for 1855, and deserves
to be well considered by curators of museums, and all
who take an interest in such institutions.

The Ipswich Museum was not the only one bene-
fited by his contributions and advice. The Botanical
Museum at Cambridge, which was got together entirely
by himself, and which would have been much extended
but for want of space, has been already spoken of.
But he was a great benefactor also to the Museum of
Economic Botany at Kew. Sir W. J. Hooker has justly
alluded in his guide-books to the important aid received from him in this Institution. Not only did he
contribute largely to its stores, by a free remittance of
the best and finest specimens he could supply of whatever came into his possession deserving a place among
its collections, but he assisted both in the general
arrangement of the whole, and in the details connected
with the operations of cutting, mounting, and ticket-
ing the specimens, so as to adapt them best for public
exhibition. The admirable way in which all this has
been executed by himself, or from his suggestions,
shows, like the Museum at Ipswich, " his unequalled
talents for such work." No extraneous aid of any
consequence, I am informed by Dr. Hooker, has been
obtained from any other source.

The tact he had for preparing museum specimens,
especially in his own department of Botany, was
further shown on the occasion of the Great Exhibition

at Paris, (following the one in London in 1851,) to which " he communicated a most beautiful series of Carpological Illustrations, which excited the enthusiasm of the Paris botanists, and of which a duplicate set is now in the South Kensington Museum."

And what he so much excelled in, he took delight in. He never seemed more thoroughly contented than when at work at his table, on a long winter's evening, with all his specimens and tools about him; now selecting the best specimen of each kind, or examining it with a lens; or, in some cases paring it with a knife in order to show the interior;—now cutting stiff card-board with strong scissors of a peculiar make he had on purpose;—now moulding soft clay into a suitable form to serve as a stand for specimens that could not otherwise be so well inspected. This last was an ingenious method he devised for mounting and exhibiting any objects, especially minerals, which if left to themselves could not be made to present to the spectator that particular face in which resided their chief peculiarity. The clay was afterwards suffered gradually to harden, " and became a firm support to the object. The cement employed was liquid glue, $i.\ e.$ shell lac dissolved in naphtha. " *
Many specimens thus mounted may be seen in the Ipswich Museum. In such work he would be quite absorbed with his family around him, taking little notice of what others were about. It was always evident that his mind was quite as much engaged as his hands. He would seem to be turning over in his

* See Report of Brit. Assoc. 1856. p. 461.

thoughts for what uses each specimen might serve, the impression it was likely to make on the spectator's eye, the instruction it was capable of affording, the desire it might kindle up to know more about it as also to become acquainted with other allied objects of similar interest.

The educational element as connected with institutions for the exhibition of specimens either of natural history or the arts was always uppermost with him. He often regretted how insufficient the labelling generally was in museums, how much more instructive it would be if pains were taken to extend the information respecting each specimen to something more than the mere name, and country from which it comes, to which in so many instances the labelling is confined. A few important facts connected with its history, affinities, structure, or uses to which it was put, according to circumstances, might be brought into a small compass, while it would greatly increase its interest with the public. This accordingly was what he always looked to in museums, more than to the absolute extent of the collections themselves. I once went with him to see the Museum of the Bristol Institution. After examining it well, he remarked, " Much has been done to render this museum useful and instructive, but it is incomplete in its arrangement : the master mind has left it." There were indeed few master minds besides his own in this particular department of natural history. Many museums have been well arranged for general display, but they have seldom yielded all the instruction they might have afforded, if the grouping and arranging had had an especial reference to this end, for which perhaps few curators have sufficient judgment and experience, or if they possess these, they are not always able to bestow the requisite time and attention to do full justice to the establishment confided to their hands.

His great desire to see education more widely extended, and especially a more general diffusion of the knowledge of the natural sciences, led him to take as much interest in schools and colleges as he took in museums. He was one of the original members of the Senate of the University of London, nominated in the Charter dated November 28, 1836, an institution which he thought well calculated to promote the study of the sciences and other branches of literature, among those classes of men, who seldom find their way to the ancient universities of Oxford and Cambridge. He acted as an examiner from the first nomination of the examiners in 1838 until 1860, a period of twenty-two years: he then resigned that office, but continued a member of the Council till the time of his death. His examination papers are highly spoken of. Dr. Hooker has been kind enough to send me the following remarks in respect of them, and his method of teaching botany in this University, which I give in his own words :—

"He was the first to introduce into the botanical examination for degrees in London the system of *practical* examination. He insisted that a knowledge of physiological botany, technical terms, minute anatomy, &c., were not subjects by which a candidate's real knowledge could be tested, for the longest memory must here win the day. Still less did it test the observing or reasoning faculties of the men. Further, any amount of book-knowledge on these subjects was consistent with the most profound ignorance of the very rudiments of botany, and of the most elementary knowledge of the parts of plants, the relations of those parts, and the mutual affinities of plants. He there-fore insisted in all his examinations that the men should dissect several specimens, describe their organs systematically, and be prepared to explain their rela-tions, uses, and significance, in a physiological and classificatory point of view; and thus prove that they had used their eyes, hands, and heads, as well as their books.

" By this it must not be supposed that he over-looked pure physiology and minute anatomy.   At every examination he gave questions in these branches, but their knowledge of them was tested by practice. In short he regarded the acquirement of an abstract knowledge of these branches as an accomplishment, which the tyro can only obtain by reading, and not of much educational value as a means of training the faculties.   He complained of the methods of teaching botany adopted in London particularly, where, espe-cially in the purely medical schools, every branch of the subject is crowded into a three months' course, and the men leave the class with a most indistinct no-tion of the very rudiments of the science, and with minds positively impaired by the cramming they had undergone, instead of strengthened, as they should have been, by a judicious use of specimens, and by confining the course to what men of average intellect can acquire *well* in the time. ' They are badly taught ' was his constant complaint: ' Botany has done them more harm than good.' They will enumerate every variety of form of cells and fibre, and give a technical name for each with perfect fluency, and yet be ignorant of the parts of a flower, fruit, or seed, of the great divisions of the vegetable kingdom; and they have not an idea how a plant grows, and in what order its parts are evolved.

" He was further one of the few who thoroughly appreciated the diverse faculties of young men. Himself good at mathematics, classics, and the biological sciences, he yet clearly understood that with the majority of men a faculty for acquiring one or more of these branches of knowledge was deficient : in some, one was over-developed at the expense of the others; and in his appreciation of the merits of candidates he bore this in mind. Still he always held that a man of *no* powers of observation was quite an exception; and that the lowest development of that faculty was capable of such further development as to fit a man for any calling. Its apparent total absence he regarded as equivalent to its totally neglected culture; and this rendered the man unfit for a scientific or learned profession. In whatever profession, then, botany was an appointed study, he regarded total ignorance of its elements (when tested by specimens) as an insuperable barrier to passing. In short, he unhesitatingly *turned* men for botany, if they were deficient, whatever their merits in other respects."

To his own University (Cambridge) he ever retained feelings of the strongest attachment. It was the place where he had received the most valuable portions of his own education, where he had first acquired a

knowledge of the exact sciences, which so much aided
him in his various pursuits in after-life,—where he
had laboured for so many years to foster a taste for his
own favourite study of natural history, and where he
had formed so many acquaintances with men of high
esteem, some eminent in science like himself. It must
have been gratifying to him to keep up a correspondence through life, as he did in some cases, with men
who first imbibed their love of natural history from
him, or who owed to his encouragements the steadiness with which they had followed it up afterwards, and
whose knowledge of botany in particular had been
mainly acquired in his lecture-room.

But it must have been a source of yet higher satisfaction to see at last the due recognition of the Natural
Sciences as deserving a place among the subjects required of the young men for examination for degrees.
He had always thought the course of study at Cambridge too restricted. He was not one of those who
decried either mathematics or classics, both which he
had himself sufficiently mastered to know well their
value and uses. Still less did he wish to interfere
with any teaching calculated to insure a "religious
education." But he thought that more consideration
should be given to some other subjects, which, though
publicly lectured on by certain of the Professors, received no sufficient encouragement from the University
to induce the young men to take them up. It was in
1846, when making his appeal to the members of the
University in behalf of a new botanic garden, that he
first openly spoke his sentiments on this head. He
says :—

" I claim to be as warm an admirer of so much of that sound learning and religious education as is upheld by us, as any other son of our Alma Mater. No one who has spent the same number of happy years within the precincts of our University as I have done, or who has experienced the same amount of kindness and friendship there, could ever wilfully cast a slur upon its residents without being grossly ungrateful for past enjoyment, and wholly unfitted for any that may yet be in store for him. But it is the very desire to see our University improving, that induces me to state what I believe to be a few plain facts, and to give expression to the opinions of one who for many years took no inactive part in the encouragement of natural history to the little extent which his opportunities allowed him."

He then speaks in reference to his own particular department of botany :—

" I hope I should be among the last to reflect upon the resident members of the Senate for any supposed backwardness about upholding the particular science for whose encouragement I am more especially bound to plead. I most fully and gratefully bear my testimony, that from the time I first occupied the botanical chair, I have ever found the Senate ready to accede to whatever proposal I have made to them, from time to time, for rendering me assistance, either in purchasing specimens for our botanical museum, or in providing the means for securing their preservation. But still I must consider the claims of botany are not sufficiently apprectated among us. There are persons of great mathematical and classical attainments, who have very erroneous notions respecting the ultimate aim and object of this science. Many persons, both within and without the Universities, suppose its objects limited to fixing names to a vast number of plants, and to describing and classing them under this or that particular ' system.' They are not aware that systematic botany is now considered· to be no more than a necessary stepping-stone to far more important departments of this science, which treat of questions of the utmost interest to the progress of human knowledge in certain other sciences which have been more generally admitted to be essential to the well-being of mankind. For instance, the most abstruse speculations on animal physiology are to be checked, enlarged, and guided by the study of vegetable physiology. Without continued advances in this latter department of botany, the progress towards perfection in general physiology must be comparatively slow and uncertain. As regards the progress of botanical physiology, even chemistry itself must be viewed as a subordinate assistant, whilst it is making us acquainted with those physical forces by which mere brute matter is regulated and arranged. Those forces are themselves to be restrained and modified by the instrumentality of vegetable life, in bodies whose appointed position is to prepare all the organic matter that is destined for the support of a still higher race of creatures in the general scheme of nature. We may feel quite confident that some of the arts we consider to be most important to man, such as agriculture and horticulture, will never be perfected until the fundamental principles of vegetable physiology shall have been satisfactorily elucidated.

Numerous indeed are the bearings, direct and indirect,
which botany holds upon other sciences, and upon
various arts!   And, here, I cannot refrain from
making some allusion to the station which botany
might be made to occupy in an improved scheme
of liberal education.   But as this claim has been so
fully recognized by Dr. Whewell, in his late publication ' On Cambridge Studies,' I shall prefer reminding you of his opinion, to urging upon you whatever
may be my own view."

He then quotes passages from the above work, in
which Dr. Whewell sets forth the grounds upon which
he "should much desire to see botany, or some other
branch of natural history, or natural history in
general, introduced as a common element into our
higher education, and recommended to the study of
those who desire to have any clear view of the nature
of the progressive sciences; since it is, in fact, the
key and groundwork of a large portion of those
sciences."

After which Professor Henslow proceeds as follows :—

" Very possibly, as some suppose, the time may not
yet have arrived for these sentiments to meet with any
loud response among us ; and certainly I am not to be
considered a competent judge on this point.   Still I
trust you will allow me to remind you that we are
very often blamed by persons who do not understand
the peculiarities of our establishment, for not doing
more than we at present attempt for the natural
sciences.   All know that to a limited extent some of
them are encouraged ; but they know also that not

one of them is patronized, like mathematics and
classics, by the stimulus of rewards for securing its
diligent and successful cultivation. What Dr. Whewell
has said at p. 224,* concerning the establishment of
a general Tripos for students in the progressive
sciences, appears to me well worthy of our most
serious attention.   Every one who has had much experience in preparing pupils for an ordinary degree can
bear testimony to the facts of there being minds naturally incapacitated for clearly and fully comprehending
a mathematical problem.   Now there are many persons
with this want of mathematical ability, who will de-
lightedly occupy themselves in one or other department of the natural sciences.   It has long appeared to
me a mistaken policy to bind down such persons to
one particular routine of dull anxious plodding, without allowing them the opportunity, before quitting the
University, of proving that they have not laboured
unsuccessfully in the general field of human knowledge.   But I shall not dwell upon the various and
numerous considerations which incline me to accept
the view that has been already taken by Dr. Whewell,
of the propriety of our adding botany, or some other
branch of natural history, to the general curriculum of
a liberal education."

These sentiments, though in agreement with what
had been previously advanced by so high an authority
as Dr. Whewell, were not immediately taken up by the
University.   "Great bodies move slowly," especially

* Alluding to the work above referred to—"Of a Liberal
Education in general ; and with particular reference to the
leading Studies of the University of Cambridge," 1845.

if at the same time ancient institutions; and it is seldom that they can be induced, till after many stirring appeals, to alter their ways, or to admit of any innovation in their existing practices. Nevertheless, within two years after the publication of Professor Henslow's pamphlet, the matter to which he adverted came to be more seriously entertained, and an attempt made to introduce the desired change. The news reached him whilst he was preparing his " Address " for the Ipswich Museum before spoken of, and in which he alludes to it. He first mentions a letter he had received from a correspondent, who made the same complaint respecting the neglect of the natural sciences at Oxford. The letter states,—" Mr. *** is an extremely agreeable and intelligent man in everything but natural history and science, of which he has a lamentably low opinion. He is a perfect specimen of the too partial system of education pursued at Oxford; and as ignorant of the origin and working of our most common manufacturing products and arts, as he is well informed on all matters of finance, policy, &c." Upon which Professor Henslow observes :—

" I fully agree with my scientific friend, that our Universities have been unduly negligent of the claims of science; but I am truly happy to find, from a printed communication just received, that an effort is being made at Cambridge in favour of adding some of the natural sciences to the hitherto exclusive system which has prevailed there. I am informed that a good deal of opposition is expected; but the attempt is considered worth making if it shall only show the world that there are persons in the University desirous of pro-

I

gress. For myself, I consider the attempt well worth making, if it shall only show us whether or not there is any prospect of a beneficial change being made in this direction."

In full hope that the day was not far distant when he would see his own favourite science taken into the number of those subjects which would be required of all candidates for a common degree, he lost no time in preparing and publishing a new "Syllabus of a Course of Lectures on Botany, suggesting matter for a Pass-Examination in this subject." It was "very hurriedly put together," and intended " as no more than a rough approximation to what he considered might be reasonably expected of men whose time might have been pre-occupied in mastering the subjects already required" for that purpose. The Syllabus is divided into the three sections of structural, systematic, and physiological botany, each of which branches of the science he thought ought to be considered in drawing up an examination paper. He imagined it might "be found sufficient to restrict the questions of a pass-examination to the following topics :—

" 1. Definitions and explanations of a given number (say about 200) of the most essential technical terms—a list of which may be inserted in an Appendix.

" 2. Diagnoses of the British families which have been selected under section 2. These should be learnt by comparing plants or figures with the descriptions that are given in systematic works, and ought not merely to be got up by rote.

" 3. Any specimen, figure, or description of a plant belonging to one or other of these British families, should be referred to the position it would *strictly* seem to occupy (upon inspection) in the systems of Linnæus and Jussieu, so far as they are set forth in the tabular view given in this Syllabus.

"4. A selection from the physiological topics discussed under section 3."

The above sketch had no reference to "Candidates for Honours" in a Natural Sciences Tripos, in the event of such a Tripos ever being established by the University, which he probably at that time thought very doubtful, but simply to "the (so-called) non-reading portion of the undergraduates," who were desirous of taking a common degree.

The day, however, at length arrived, when he saw more than he at first ventured to hope for, in respect of a required examination in these sciences, fully carried out. The "opposition" that was expected, and which was at first shown, ultimately gave way; though it took some time to secure the success of the new measures, when they should be laid before the Senate; and it was by very gradual steps that the present system came into operation, only receiving its full development a short time before his death.

The first step agreed to by the Senate, was to require of every candidate for an ordinary degree, in addition to what was formerly required of him, that he shall have attended the lectures of one or more specified Professors, and shall have obtained a certificate of having passed an examination satisfactory to one of the Professors whose lectures he has chosen to attend. Among these Professors was that of Botany;

but as it was left to the young men to make their own choice among eight Professors, the new rule did not necessarily augment the number of botanical students. The Grace for this new regulation passed Oct. 31, 1848, being the autumn after the delivery of the Ipswich Museum Address, when he mentioned having received the first intimation of the subject being taken up.

The next step was the institution of a Natural Sciences Tripos for such men as, having offered themselves for honours, should pass the examination in the subjects to which that Tripos relates; a place in this Tripos, however, not conferring a degree, which was still only to be obtained, and which it was necessary previously to have obtained, or at least passed the examination for, in the usual way. This Tripos first came into operation in 1851. The same year Professor Henslow published his "Questions on the Subject-matter of Sixteen Lectures in Botany, required for a Pass-Examination," in which he endeavoured to fix, more definitely than he had done in the "Syllabus" of 1848, the amount of what might be fairly required for such purpose, the questions having reference to the course of lectures he had just been delivering, a reference being also given under each question to the page of the syllabus bearing upon that question.

At last it was decided by the Senate, on February 23, 1860, "that all students who shall pass with credit the examination for the 'Natural Scieuces Tripos,'" having passed the previous examination, and what are called the "Additional Subjects" (an examination in which is required of all candidates for honours), " be entitled to admission to the degree of Bachelor of Arts."

The first year in which it was thus possible for men to get a degree by passing in honours in the Natural Sciences Tripos, was 1861, being ten years after the institution of the Tripos itself. And one of the last acts of Professor Henslow's life,—his very last as connected with his Professorship and the University,— was to examine the candidates for honours in this Tripos, in his own department of Botany. The examination commenced on the 4th of March, and lasted six days. The Tripos list appeared within a few days after the conclusion of the examination, containing the names of three men, one in each of the three classes, who were admitted to a degree accordingly.

Thus at length, only a few weeks before his death, he saw the measure, which he and others had at heart, fully established, and a great part of the reproach wiped away, which had so long attached itself to the University, for not giving more encouragement to the study of Natural History.* It may still perhaps be said, that it has no Professorship of Zoology, in which department Oxford has lately taken the lead. But it should be remembered, that but very few of the Professorships in Cambridge have been founded

* "In the Christmas week" (of 1843) the late Professor Edward Forbes paid a visit of two days to Cambridge, and inspected the colleges with Mr. Ansted and Mr. Babington. "I was greatly pleased with my visit," he says, " except in one thing,—to find that natural history is discouraged as much as possible, and regarded as idle trifling by the thousand and one mathematicians of that venerated University."—*Life of Edward Forbes*, p. 355.

by the University itself, most of them being the
foundations of Benefactors; and that where they have
been so founded, it has not generally been till after
the delivery of several courses of lectures on a particular subject by some person who has come forward
for the purpose, having the University's sanction, and
upon whom, in consideration of such lectures, that
body has afterwards conferred the title of Professor.
Even then, however, supposing that any one had leave
to deliver lectures on zoology, and the lectures were
approved of, it might be difficult, perhaps impossible,
for the University to promise any salary to such individual, as Professor, under its present circumstances,
when it has been recently raising considerably the
salaries of some other Professorships, and there might
be no further funds available for such purposes. Until,
therefore, an offer has been made by some private individual to found and endow a Professorship in that
branch of science, and the University decline the offer,
it is, perhaps, hardly fair to throw blame upon the
latter for its apparent neglect of the subject in question. In the mean time, " Comparative Anatomy and
Physiology with Zoology," are duly treated of in the
course of lectures delivered by the Professor of Anatomy, while many of the questions proposed to the
candidates for honours in the Natural Sciences Tripos,
in the Examination Papers of 1861, are as purely
zoological, and require as much knowledge of that
branch of Natural History on the part of the student
to be satisfactorily answered, as could be reasonably
desired.

None of the Graces connected either with the introduction of the Natural Sciences among the subjects required for an ordinary degree, or with the establishment of a Natural Sciences Tripos, originated with Professor Henslow. Any such active interference, from his non-residence in the University, was hardly practicable. But he had the gratification of reflecting that the final adoption of these measures was in unison with the sentiments he had always uttered. He could look back with complacency to the time when, being resident, he had used his utmost influence to draw attention to the importance of Natural History, ever setting forth its claims and attractions,—promoting the study of it among the young men especially, and aiding by his teaching and example all who were induced to take it up. It can hardly be thought improbable, that the earnest endeavours he then made to diffuse a taste for his own favourite pursuits, may have had their effect in leading to this altered state of things. They may have served to pave the way for the new regulations, by turning other men's minds besides his own, to the subject. When we consider the utter disregard paid to Natural History in the University, previous to his taking up his residence there,—and that, though a Botanical Professorship had long existed, there had been, as before stated, no lectures for thirty years, even in that branch of the science, some knowledge of which is generally considered essential for at least medical students,—we can hardly believe that the Natural Sciences would at this day have so far found a resting-place in the University as to have a Tripos to themselves, had the same feelings formerly entertained respecting them

prevailed uninterruptedly to the present time. Yet it
would be difficult to fix upon any other individual
than Professor Henslow, who would have exercised
the same powerful influence in behalf of the claims
of Natural History, or awakened the same attention
to it among the members of the University generally.

Besides their importance in respect of the actual
knowledge they convey, and the insight they give us
into " some of the physical laws by which the works
of the creation have been advanced to their present
condition," Professor Henslow advocated the value of
the Natural Sciences as assisting the inquiries of those
learned men who devote themselves mainly to the
study of the higher truths of Revelation. He considered that a knowledge of " verbal criticism," however
deep and extensive, was not in itself sufficient to
enable persons to " catch the spirit of the Scriptures."
He spoke of " bygone superstitions," exploded by
the progress of science, which hindered " the best exertions of some of the foremost and most devoted
champions of the truth in former generations ; " and
there were " certain minds," in his estimation, even at
the present day, " deeply imbued with scriptural learning," but not altogether devoid of prejudices and
errors, arising out of their " partial disbelief in some
of the best-established truths which our improved
acquaintance with God's works has at length revealed
to us." To guard against such mistakes, he thought
it " well worthy consideration, whether a University
can safely cease from esteeming it a duty to encourage
any department of sound learning which may assist
in improving our general views of God's Providence.

Whatever might have been excusable in ages of comparative ignorance, surely our present wisdom is to accept of every improved light we have now obtained from the works of God, as well as from every learnedly-corrected interpretation of His Word: so that these two witnesses to His infinite power, wisdom, and goodness, may mutually corroborate each other." There was no place, in his opinion, "in which all unworthy notions about the results to which the Natural Sciences may yet conduct us, ought to meet with more decided contradiction and rebuke, than within the walls of a University. And yet (he asks) how is this likely to happen again, if the tendency to neglect these sciences (I do not say to banish them) shall continue to increase, as most certainly it has of late years been doing among us?"

The above sentiments were expressed in 1846, in his pamphlet on the Botanic Garden. His joy and satisfaction at finding that the Natural Sciences had been at length duly taken up and patronized by the University, will appear in the following passage at the end of the Preface to his "Questions, &c., for a Pass-Examination," written in 1851; in which he speaks of the unalloyed pleasure the study of these sciences is calculated to afford, while at the same time he again alludes to their importance in a religious point of view :—

"Many will rejoice in after-life, that the University has required them to pass an examination in one or other of the subjects treated in our Professorial lectures. By mastering, whilst at college, the little difficulties which stand in the way of all who wish to

learn 'how to observe' the works of nature, they will
have obtained a sufficient grasp of such subjects to
make them afterwards agreeable as well as instructive
occupations.   If they will trust the experience of one
who has passed the last twelve years in an out-of-the-
way purely agricultural country village, where even
many a learned classic or profound mathematician
might very possibly have felt himself occasionally at a
loss for want of further intellectual companionship
than books alone can afford, I can assure them there
is nothing like possessing the power of conversing
with the works of nature for a constantly unalloyed
source of enjoyment at all seasons of the year.   However the wit or the wisdom of man may delight us,
the excellency of God's works leaves them far behind.
I may be presumptuous in making the remark, and
many will esteem it ill-advised; but it has often
appeared to me, that the study of God's Word alone
may not always be sufficient to protect some of us
against the morbid imaginings of these uncertain
days; and I therefore the more rejoice to find the
Natural Sciences at length taking firmer root in this
seat of sound learning and religious education, than
they have hitherto obtained."

CHAPTER VII.

MISCELLANEOUS OBSERVATIONS IN NATURAL HISTORY.

PROFESSOR Henslow was so much taken up with educational work, and the various schemes he set on foot for the benefit of his parishioners, besides attending to their spiritual concerns,—with the duties also of his Professorship—that he had not much time for authorship, after taking up his residence at Hitcham, beyond short communications to different scientific and other journals. But though he published few original researches, he was full of original ideas, the fruits of a patient and careful observation, extended over all parts of nature, and carried on through every period of life. Wherever he was, and whatever company he might be thrown into, he let slip no opportunity of getting all that was to be had by looking for, or inquiring after. He seldom went from home, or even took a walk into the village, without gleaning some new fact, or adding some new specimen to his collection, any especially that tended to illustrate his lectures in the University, or his more popular addresses to his own people. He had a singular tact for seeing at once what there was distinctive or remarkable in the different phenomena that came under his eye,—for laying hold of what there was in them that had any bearing upon other subjects of inquiry, or that could be turned to any account in furthering scientific truth. Things that would be thrown away by most people, and which were really useless, from decay or injury, in respect of the purposes for which they were originally kept, he would sometimes make do service in another way by illustrating, it might be, the changes of form or structure through which they had passed or were passing. Thus, on looking over a collection of shells, many of which had been entirely spoilt by having been kept several years in a damp apartment, he selected some specimens of the well-known *Ianthina*, once fine and beautiful with their rich violet hues, but then distorted in form, covered with a white flocculent powder, interspersed with minute acicular crystals,—and these he at once put aside as exemplifying the manner in which the gradual decomposition of shells takes place. And so, if any strange-looking natural object were put into his hands, he would, in most cases, quickly notice to what its peculiarity was owing, and what was to be learnt from it.

If, on the other hand, he met with anything the exact nature of which he did not quite understand, and he had reason to think there was some hidden truth worth seeking for, he would never rest till he had elicited it.

Hugh Miller observes, that " sticks and stones have a story to tell; " and no man would more

thoroughly get the story out than Professor Henslow.
There was a determination and steadfastness of purpose about him, which was the great secret of success in all that he undertook. This was a remarkable
feature in his character. It showed itself alike in his
investigations of science, in the plans he organized for
the improvement of his parish,—in any matter of less
moment, even to a child's puzzle, to which he had
once been induced to give his attention. The more
perplexing it was to account for anything, the more
he would turn the matter over in his mind till the
right explanation offered itself. Questions that were
put to him which he could not answer immediately
were not forgotten, but generally had their answer
afterwards, when perhaps they had been forgotten by
those who put them. He would leave no stone un-
turned,—no means that afforded a chance of success
unavailed of, to obtain what he was in search of.

It was this turn of mind, generated by an innate
love of nature, fostered and improved by long experience and quickness of perception, that gave him a
habit of looking to sources of information on scientific
subjects that many would have passed over or thought
barren of any probable results. There were some subjects, not connected with his own department of
botany, in which he particularly interested himself.
One of these was the formation of the various kinds
of pebbles and agates, and the origin of the bands of
colour, more or less parallel, or irregularly circular and
concentric, by which they are so often marked. So long
back as November, 1836, he made a communication to
the Cambridge Philosophical Society, offering conjec-

tures on this subject, but it was merely oral, and he never published the result of his inquiries, which he continued to renew at intervals through life. He left behind him a few rough notes and sketches relating to it, but too imperfectly put together to form a sufficiently just notion of his views, so as to render it safe to give full publication to them. The general theory upon which they rested seems to be, that "if a defined mineral mass (fragmentary or nodular or crystalline) is embedded under circumstances which admit of its being continuously acted upon by moisture or heat, any colouring matters introduced, or originally included in its substance, can re-arrange themselves into zoned laminæ, which bear a distinct relation to the *surface* of the mass. Also nodular aggregations can be formed within its substance from numerous centres, many of which are seated at or near the surface of the mass."

He had formed a large collection of pebbles in illustration of the above subject, many sliced and polished to show their internal structure, and he was often adding to it. These specimens were obtained from various quarters; but not a few were picked up in his own neighbourhood. In the search for any that might tend to confirm the views he had been led to entertain, he would occasionally stop long, with hammer in hand, to the great wonderment of passers-by, to examine the stones that lay collected in heaps for mending the roads, speculating at the same time on the sources whence they had been derived, and on the long distances many had travelled over, before finding a resting-place in the Hitcham fields.

In like manner he was always on the look-out for

facts in botany. This, indeed, was his own particular field, every part of which he most thoroughly explored. Structural botany, however, normal and abnormal, had an especial attraction for him. He would notice in his walks any little variation of character in the weeds by the road-side, that helped to confirm generally-received views respecting their organization and affinities, any monstrosities of leaves and flowers that served to establish the laws of vegetable morphology. The extensive collection that he had made of such specimens, along with others in which there was some abnormal growth arising from disease, or the attacks of insects, has been before alluded to, when speaking of the Cambridge Botanical Museum.

I have said in an earlier part of this memoir that he never entirely abandoned mineralogy; and the colouring matter in pebbles above noticed, was one of those questions in the science which he was always thinking over in his own mind. Neither did he ever entirely abandon zoology, which was the first branch of Natural History he ever took up, and to which he equally devoted as much time as other more important occupations would allow. It is astonishing to reflect on the multitude of little facts he had got together, and the extent of his collections, in this department of science; quadrupeds, birds, reptiles, shells and insects, all alike finding a place in his overflowing museum, to which nothing ever came amiss. His eye was quick in detecting new species which he had not before observed in his own neighbourhood, and even when considerably advanced in years, he would, with a zeal and alacrity equalling that of the youngest entomologists, pursue and capture any rare insects that came in his way.

Many such rarities were to be seen in his cabinet, including that fine butterfly, so irregular in its appearance, the Camberwell Beauty, * which he took in his own garden.

But he did not content himself with mere collecting; he was a close observer also of the habits of animals. He kept alive for some years two large Jersey Toads, which furnished him with matter for two communications to the *Gardener's Chronicle,* respecting the way in which they shed their skins,—written in his own peculiar humorous style. † The object of the second communication was to point out a circumstance attending this operation in one instance different from what had occurred in a previous one; viz. the toad did not eat up his own skin afterwards, as toads have generally been observed to do; upon which he judiciously remarks,—" I send the account, because it shows how necessary it is to multiply observations before we can arrive at sure conclusions respecting the habits of animals, and the extent to which they may become varied by circumstances."

There was a tribe of insects, in which he was particularly interested, and to the habits of which he had paid the closest attention; viz., that of wasps, hornets, and humble-bees. He had studied the construction of their nests, and made remarks on the economy of some of the parasitic larvæ found in them, which form the subject of a communication to the *Zoologist* ‡ which must not be passed over. He had " often amused himself with taking the nests of those insects,"

* *Vanessa antiopa.*
† *Gard. Chron.,* 1850, pp. 373 and 422.
‡ 1849, vol. vii., p. 2584.

specimens of which he was desirous of preparing "for the Museum at Kew, his friend Sir W. Hooker seeming to think these products of our 'native papermakers' would be legitimate appendages to his most interesting illustrations of economic botany." In the article above referred to, he has first described the ingenious way in which he succeeded in taking their nests without being stung. "A large hornets' nest" had been found "attached to the gable end of a cottage," another "in the middle of a hollow tree."

"This latter (he says) was brought home on the 18th of October, when I discharged from it 166 queens, 22 drones, and 9 neuters; and on the same day I discharged from the large nest, above alluded to, nearly 100 queens, a few drones, but no neuters. The nests were on my study table, and the insects were compelled to evacuate them by my pouring alcohol over them : as they came out I caught them with a pair of dissecting forceps, and kept many of them alive for some months under a large bell-glass. The neuters soon died, the drones did not live long, but the last of the queens was alive as late as February. A little attention to their habit of flying from the nest directly to the window preserved me from being stung at home; and neither myself nor assistants were stung abroad, though we took more than six hornet-nests, with no other precaution than the active use of a butterfly-net, one person working whilst the other stood guard. They often flew directly at us, but by standing perfectly still, and gently waving the net, they were always persuaded to change their aim, and were caught and killed accordingly. Owing to their building in exposed situations,

we found it impossible to stupefy or kill them with
spirits of turpentine, as we can so easily contrive to do
with the wasps; and they have the awkward habit,
moreover, of being very active all night, running about
the tree in which their nest is lodged, and flying
directly at a lantern which may be in the hand of
the too curious observer."

He then proceeds to speak of the parasitic larvæ
found in these nests, some of which he bred. One
was the *Velleius dilatatus*, a beetle so rare that Mr.
Stevens, in his "Illustrations of British Entomology,"
states that he believed his own specimen to be the only
example up to that time captured in England. Of this
insect he found "thirty or forty specimens" in the
hornets' nest taken from the hollow tree, which, however, he did not succeed in rearing to the perfect state:
they were observed "actively traversing the comb, and
poking their heads into the cells in search of food,"
which was presumed to be the larvæ of the hornet.
Other parasites, the larvæ of insects better known, are
then described, from the nests of wasps and humble-
bees. Some of these, "the larvæ of certain parasitic
Diptera," or flies, deserve notice in connection with a
curious little fact in the construction of wasps' nests,
first observed, I believe, by Professor Henslow. He
says, "These larvæ, at least in their early stages, are
found crawling about the loose stones and earth immediately below the nests of wasps. When the wasps
(*Vespa vulgaris* and *V. rufa*) excavate the cavity in
which their nests are built, they are unable to remove
large stones, which continue to subside as the excava-
tion advances, and, ultimately, form a sort of rude

pavement below. I have placed an example or two of this in the Kew Museum. These stones are kept moistened by matter dropping from the nest, and on this and dead wasps, the maggots seem to revel."

Several other curious little observations, which show the closeness with which he looked into nature, are recorded in the above short memoir, to which the reader who wishes to know more of the habits of these insects is referred. Specimens of some of these nests, including the splendid hornets' nest attached to the gable end of a cottage above alluded to, may be seen in the Museum at Kew, more perfect of their kind, and more admirably mounted for public exhibition, than any perhaps to be found in other similar institutions.

In reference to the method of taking wasps' nests with spirits of turpentine, spoken of above, it may be added that this was another ingenious contrivance first hit upon by himself, and described by him in the *Gardener's Chronicle* for 1842.* The method consists of simply " pouring about half a cupful of spirits of turpentine into and about the entrance-holes " after dark, when the wasps, with the exception perhaps of a few stragglers, are all in for the night : " then place a flower-pan over it, and bank it round with earth." This has the effect of stupefying them ; and, if desired, the nest may be dug up thirty-six hours afterwards with perfect security.

Another minute observation of his is recorded in the *Zoologist,* respecting the food of micro-lepidoptera, or small moths.† His " attention was attracted to a small caterpillar feeding among the flowers of *Pim-*

* P. 637.   † *Zoologist,* vol. x., p. 3358.

pinella saxifraga, in Hitcham parish." After carefully
searching he found four others of the same kind; he
" placed them in a glass jar, and fed them for two or
three weeks," when he " observed that the petals were
gnawed off and lay scattered about, and that only the
innermost parts of the flowers and summits of their
pedicels were eaten." Four of these insects he reared
to the perfect state, and they proved to be an uncommon
and apparently unnamed species of *Eupithecia.* In the
case of another species of this genus, *E. linariata,*
which is described in books in a general way as feeding
on *Linaria vulgaris,* he observed that the caterpillars
" bore into the unripe capsules in order to feed on the
young seeds."

The above little facts, deemed worthy of record by
Professor Henslow, and a mere selection from the many
he noticed, show how diligently he used his eyes in
searching into the ways and habits of animal life, and
how active he was in his inquiries, even when going
the rounds of his parish, and when his mind must have
been often turning at the same time upon more serious
matters. They afford a lesson to young observers, and
to all persons circumstanced as he was (to use his own
words before quoted), " in an out-of-the-way country
village," without many resources of amusement besides
those which they can find for themselves in such or
similar pursuits. Even the most unattractive spots
may be sometimes made to yield the richest fruits if
they are only carefully sought after. We should never
consider that we have gleaned all that is to be got: the
same ground may be gone over year after year without
exhausting its natural productions. It should be re-

membered that the highest generalizations in science are based upon accumulated facts, which few have it not in their power to add to, if they are so disposed. In the Natural History sciences, success depends mainly upon the habit of never passing anything over as trifling or common, unless we are quite sure we know all about it. In reference to this last point, Professor Henslow himself has made some excellent remarks " on the Registration of Facts tending to illustrate Questions of Scientific Interest," * intended more especially for travellers and others who are desirous of collecting for naturalists without having themselves much knowledge of Natural History. Such persons are too apt to have their attention exclusively directed to pretty or peculiarly striking things, overlooking *common* things, which oftentimes have the most interest in the eyes of those who really understand the subject. He says :—

" The fact is, that persons who are no naturalists are no judges of what objects are most likely to be of interest in a strictly scientific point of view. Botanists would rather receive one of our most common weeds from a newly-discovered, or newly-explored country, than a new species of an already known genus. There are higher departments of Botany than mere collectors of specimens are aware of : to ascertain the geographical distribution of a well-known species is a point of vastly superior interest to the mere acquisition of a rare specimen."

One feature which particularly distinguished many of the researches and observations made by him in Natural History was their practical value. Though he

* *Gard. Chron.*, 1844, p. 659.

loved science for its own sake, he was never more
pleased than when he could turn it to useful purposes. We have already seen how ready he was to
assist the farmers in their agricultural operations, by
suggesting and helping them to make experiments
upon manures, &c.; and he was equally ready to ascertain the cause of mischief done to crops by the attacks
of insects and other enemies, or of failure due, as it
sometimes was in his opinion, to farming upon wrong
principles. These latter researches gave occasion to
some valuable papers in the " Journal of the Royal
Agricultural Society of England," the first of which,
published in 1841, is a " Report on the Diseases of
Wheat,"* and observed to me by Mr. Berkeley, a high
authority on the subject, to be " one of the clearest
and most interesting memoirs he ever read." The
great extent to which the crops in this country suffer
from the prevalence of these diseases in certain seasons
is well known to every practical agriculturist, and
justifies any attempt to throw further light upon their
history. Practical men themselves, however, are seldom competent to undertake the inquiry. This was
shown on the occasion of the Royal Agricultural
Society offering a prize for the best essay on the
subject, when no essays were sent in considered
worthy of one. Professor Henslow saw these essays,
and it was evident to him " that the authors were ignorant of many facts long known to scientific inquirers,
respecting the nature of these diseases, and the causes
producing them." It was this, accordingly, which
induced him to undertake the report in question. It

* *Journ. Roy. Ag. Soc.*, 1841, vol. ii., part 1.

was " not written as a scientific communication, but solely with the desire of directing the attention of practical agriculturists to what had already been done on the subject, and to point out what they themselves are required to do towards advancing our knowledge of the diseases in corn, and the modes of providing remedies against them."

After some " Remarks on Parasitic Fungi," to which the greater proportion of these diseases are due, he gives the results of his examination of " wheat infested by five species of these parasitic fungi; by the ergot; by the little animalcule which produces the ear-cockle or peppercorn; and by the wheat-midge." He has entered at some length into the history of these different diseases, describing the general structure of the fungi, when occasioned by fungi,—stated the circumstances and conditions under which they usually appear,—how they may be distinguished by a common eye, and how they appear to the microscopic observer,—what parts of the plants they attack, and the mischief that arises in each respective case. He speaks of the incalculable number of sporules or seeds, in the case of the bunt-fungus,—amounting probably to many millions—which are contained in a single grain of wheat infected by this disease; whereby we may form some idea of the extreme facility with which it may spread over a whole district, if the conditions are favourable for its development. It may, however, be kept off to a great extent by steeping the seed wheat in certain solutions, a precaution now generally adopted by farmers, and which, if universally practised, might lead eventually almost to its extirpation, from the bunt-fungus having " hitherto been met with only in the grains of wheat,"* and from its seeming " to be a condition essential to its propagation, that it should be introduced into the plant during the early stages of its growth, and that its sporules are most readily absorbed by the root during the germination of the seed from which the plant has sprung."

But this process can never be effectual in the case of some of the other diseases of corn caused by fungi. Smut, rust, and mildew being all " found in the grasses which grow in pastures and by the roadside," as well as in corn, " a plentiful supply of sporules will always be kept up, to warrant our believing that we shall never expunge these fungi from the British Flora." Still precautionary measures may materially lessen an evil which cannot be wholly avoided. With regard to smut, since the sporules of the fungus producing this disease, as in the case of bunt, " enter the plants they attack by absorption at the roots, and since they are buried with those seeds to whose surface they have attached themselves, it is evident," he observes, " that too great care cannot be bestowed in procuring clean seed, or in purifying such as may accidentally be infected ; " and it has always appeared to him strange " that practical agriculturists are accustomed to pay so little attention to the raising of pure seed crops." With respect to the rust and mildew fungi which attack the stem, leaves, and chaff, there is some uncertainty whether the sporules " are

---

* Mr. Berkeley informs me that it has occurred in Algeria in a species of Hordeum. In a few rare cases it has occurred in the stems as well as in the grains.

absorbed by the roots of corn, or whether (which seems to be the more prevalent idea), they enter through those minute pores on the stem and leaves which botanists term *stomata.*" * Our information is also unsatisfactory as to other circumstances connected with these diseases, and the conditions most favourable for their development. Hence, it is not easy to suggest a cure. " Whether remedial or palliative measures may not be discovered," he considers, " is an inquiry well worthy the attention of agriculturists."

In the above report he had " thrown out a con- jecture that rust, or red-gum, and mildew, are possibly produced by two or three forms of spore generally considered as distinct, but in reality belonging to only one species of parasitic fungus." Subsequently, after further and close observation, he fully established the identity of these three fungi, and made it the subject of a second communication to the Royal Agricultural Society, published shortly after the first.† This dis- covery has been already alluded to in a former part of this memoir, in which there are some remarks by Mr. Berkeley on the subject.‡ In the paper in question, he has described and figured, with great accuracy and minuteness of detail, several of the intermediate forms ; and, independently of the fact of the above three fungi

* Mr. Berkeley sends me the following note :—" Our notions as to the mode of propagation of these fungi have been somewhat modified by the very recent discoveries of Tulasne, which show that propagation is not induced immediately from the so-called spores, but from bodies produced on their germinating threads. These facts do not, however, materially affect the statements contained in the Professor's papers."

† *Journ. Roy. Ag. Soc.*, 1841, vol. ii., part 2.
‡ See *ante*, p. 56.

K

being proved to be the same, there is the more general
truth established, that two forms of fruit may exist in
the same fungus, which he was one of the first to
ascertain. He observes that " although these details
are perhaps more strictly suited to a botanical than to
an agricultural journal, they are not entirely foreign
to the objects which the Royal Agricultural Society
have proposed to themselves—the union of science and
practice. Whatever interest may be attached, in a
scientific point of view, to determining the specific
identity of certain parasitic fungi which had previously
been considered to belong to distinct genera, the practical agriculturist must clearly be interested in learning
that two of the most fatal diseases to which his wheat
crop is liable are, in fact, only modifications of one and
the same disorder. He may thus reasonably hope
that if he is ever able to find a remedy or palliative
for one of the states of this disease, he will not need
to search further for any different corrective of the
other state."

Before proceeding to speak of a third communication by Professor Henslow to the " Journal of the
Royal Agricultural Society," it will be right to return
to his first Report, in order to notice his remarks " On
Ergot," which he considers as " certainly the most
extraordinary of all the diseases to which corn, or
indeed any other plant, is subject, from the strange
effects it produces on the animal economy." At the
time when he wrote this memoir it was a matter of
great uncertainty to what cause ergot was owing. In
itself, it is a " monstrous development of the seed of corn
and other species of the grass tribe, in which the embryo,

and particularly one part of it, is preternaturally enlarged, protrudes beyond the chaff, and often assumes
a curved form somewhat resembling a cock's spur
(from whence the name of ergot, which is of French
extraction)." Some thought this monstrosity was
occasioned by the action of a parasitic fungus, others
that it was due to the attacks of insects. Ergot, however, is now well ascertained to be a condition of the
seed induced by one or two fleshy species of *Sphæria*,
which are readily raised by keeping the ergot in a
proper state of moisture. It is of rare occurrence in
wheat, but sometimes prevails to a great extent in rye;
and, in cases in which rye bread has been eaten made
from rye infected with ergot, the consequences have
been most disastrous. The sufferers from this cause
are liable to have " their extremities rot off; and some
have been known to lose all their limbs, which, in the
progress of the disorder, fell off at the joints, before
the shapeless trunk was released from its torment. In
one instance, in France, in which a poor man, whose
family were in a state of starvation, ventured to make
bread of some ergotted rye, it killed himself, his wife,
and five out of seven of his children. Of the two
which recovered from the effects of the ergot, one
became deaf and dumb, and had one of its legs drop
off."

It is a curious circumstance that Professor Henslow,
whilst investigating the history of this particular
disease in corn, so prejudicial not merely to man, but
to the lower animals as well, should have discovered in
the parish register of Wattisham, Suffolk, a village
adjoining Hitcham, a circumstantial account of a poor

K 2

family in that village having suffered, in the year 1762,
much in a similar way to what has just been described;
and which he conceives may have been the effect of
"the family having lived for some time on bad wheat,"
which it is recorded they had done, though it does not
appear that they had used any rye. The details of
this narrative are extremely interesting, but too long
to be inserted here; and the reader who wishes to
know more of the case is referred to Professor
Henslow's memoir, in which it is given at length.
Though, as before observed, ergot has been generally
considered rare in wheat, he found it, one autumn, " in
four different fields of wheat, and gathered more than
a dozen specimens; " he found also that some of the
farmers were sufficiently acquainted with it to satisfy
him that it must be more common in wheat than has
hitherto been suspected. He concludes with " inviting
agriculturists to make inquiry in their several districts,
whether the ergot does not sometimes prevail to an
extent sufficient to induce a belief that it may be injurious to the health of the poorer classes, whose food
is little varied, and who might thus be subjected to
whatever evil influence a certain admixture of the ergot
in their flour may be capable of producing." He sug-
gests also to medical men whether some of the dis-
orders to which the poor are particularly liable, and
which are generally ascribed to poverty of diet, may
not sometimes be due to the same cause.

Professor Henslow's third memoir on the diseases of
corn contains " Observations on the Wheat-Midge." *
These observations were supplementary to what he had

* *Journ. Roy. Ag. Soc.*, 1842, vol. iii., part 1, p. 36.

before said in his Report respecting the disease in wheat caused by the attacks of this insect. The wheat-midge is a minute, two-winged fly, that " may be seen in myriads, in the early part of June, between seven and nine o'clock in the evening, flying about the wheat, for the purpose of depositing its eggs within the blossoms. From these eggs are hatched small yellow maggots, which are the caterpillars (larvæ) of this fly; and by these the mischief is occasioned." It is not exactly ascertained on what part of the flower these maggots feed; but, " in some way, they cause the non-development, or abortion of the ovary, so that the grain never advances beyond the state in which it appears at the time the flower first expands." The importance of discovering if possible some means of remedying this evil may be estimated by the state-ment of one observer, who mentions a case in Perth-shire in 1828, in which the loss, occasioned by the attacks of this insect, " in the late-sown wheats is sup-posed to have amounted to one-third of the crop!" Professor Henslow was indefatigable in his endeavours to trace the history of the wheat-midge through all its stages, but without perfect success; and it is clear that its economy must be fully worked out before we can hope effectually to lessen its numbers. It is, however, known that the caterpillars, or at least a great proportion of them, " when about to pass into the chrysalis (pupa) state, spin themselves up in a very thin and transparent web, which is often attached to a sound grain, or to the inside of one of the chaff-scales;" though it would seem " that many of the caterpillars quit the ears and fall to the ground, where (it has been supposed) they change to chrysalides, and remain buried till their final metamorphosis takes place." What causes this difference of habit in different individuals is at present a mystery. He thinks it possible that all those which fall to the ground may have been attacked by a very small ichneumon, well known to entomologists, and appointed on purpose to keep the wheat-midge in check. But, be this as it may, it is clear that, since the chrysalides of those individuals which spin up in the ear, " lie secreted during the winter among the chaff, and the fly does not make its appearance till June, multitudes of them might easily be destroyed by burning or scalding the chaff after the grain has been threshed out."

Professor Henslow stated, at the conclusion of his first Report, that it was "not written as a scientific communication;" but it is written in that masterly style which bespeaks the scientific character of its author; and the same may be said of the two subsequent memoirs just noticed. It is impossible to read all three without being impressed with the extreme care taken, not merely to ascertain what had been previously said by others on the same subject, but to examine with his own eyes the circumstances connected with the appearance of these diseases in corn, previous to suggesting experiments, to be tried in some cases, with the view of discovering the most likely means of preventing or keeping them under. He closely examined his neighbours' corn-fields, and collected a large number of specimens of the different diseases, in different stages of growth and development. When working out the history of the wheat-midge, he at-

tended the barns while the wheat was being threshed,
to ascertain the numbers of the larvæ that might be
found on the floor of the barn afterwards; then, next,
when the wheat was being dressed, to determine what
proportion of them fell through the wire-gauze, together with the dross, and what proportion were blown
out with the chaff. Nor was this all; he caused the
chaff to be sifted, as it fell close to the machine, or at
the distance of one, two, or three yards from the same,
and took the trouble to count, in each separate case,
the number of larvæ that passed through with the
dust, a certain measure of the chaff-dust being sifted
for the purpose, as well "in order to form something
like a definite notion of the numbers of the larvæ
which are housed with the wheat," as to fix the limit
of distance, determined to be three yards, to which
the larvæ are blown out with the chaff, and beyond
which, therefore, it would be unnecessary to sift the
chaff in order to collect the dust to be burnt. This
painstaking inquiry, repeated in two or three barns, is
one which few, perhaps, besides himself would have had
the patience to conduct, even if they had had all the
leisure required for it, instead of the many other avoca-
tions to attend to which drew so largely upon Professor
Henslow's time.

In 1844, we find him again assisting the farmers by
endeavouring to ascertain the cause of the failure of
the red clover crop, which was very general that particular season, and was said to have been increasing of
late years in that part of Suffolk. He thought, in this
instance, the evil might possibly arise from a wrong
system of farming. He gave much attention to the

subject, looked closely into the state of the crops, and
then published the result of his inquiries in the *Gardener's Chronicle.** The notions of the farmers were
that there was "something" in the soil necessary for
the successful cultivation of clover which had gradually
been abstracted from it, or that the failure was owing
to the attacks of insects. His own observations led
him rather to think that it was due to physiological,
rather than to chemical or entomological causes, and
consequent upon the conditions under which red clover
is universally grown, viz., "sown on land which is
simultaneously occupied by barley, or by some other
white crop." This arrangement must necessarily have
the effect of shutting out the clover from due supplies
of light and air, which are so essential to a healthy
condition of the plant. He then recommended experiments to be made to see the results of sowing clover
(intended for seed) under altered circumstances, such as
he has pointed out. In the above idea, however, he
was, perhaps, mistaken; as, in a subsequent communication to the same journal several years afterwards,† he
published some correspondence he had had with practical men upon the subject, from which he inferred
that clover "being sown with barley is not always (if
it ever be) a cause of those failures of which agriculturists frequently complain." Still, it shows the activity of his mind, and how fertile it was in suggesting
causes, which, whether the right ones or not in some
cases, ought to be considered in order to insure successful farming, but which farmers, following only their
routine practice, too often entirely overlook.

* *Gard. Chron.*, 1844, p. 529.   † Id., 1851, p. 764.

But the greatest boon he ever conferred on farmers was the discovery of the phosphatic nodules, or so-called coprolites, as applicable to manures. These nodules " abound in the tertiary formations of the eastern counties," and were known before to geologists, but it was his scientific acumen that enabled him first to see the use they might be put to in agriculture. It was "in October, 1843, that he called attention to the occurrence of phosphate of lime in pebbly beds of the red crag at Felixstow, in Suffolk; these nodules, though extremely hard, presented external indications of an animal origin, and yielded, upon analysis, 56 per cent. of phosphate of lime." From this circumstance he was led, in 1845, to think that they might take the place of bones, from which the phosphate of lime used in agriculture had, up to that time, been obtained, but of which the supply had of late years become insufficient. "The crag nodules were so abundant, that a gentleman had obtained two tons of them." Similar nodules were subsequently obtained by others from the London clay in the vicinity of Euston Square, and in great abundance from the same bed at Colchester. About the same amount of phosphate of lime was found in the London clay nodules as in the crag nodules, both of which Mr. Potter was the first to analyse, though it is believed that Professor Henslow was the first to see their value for the purposes above mentioned. The crag nodules he originally " considered to be of coprolitic origin;" but it was afterwards " satisfactorily shown that they were detrital materials from the London clay." They only differ from the nodules of this last bed in having been " slightly rolled and somewhat modified by having had a portion of iron pyrites converted to oxide of iron." "The materials which had furnished the phosphate of lime to the crag nodules appear to have been derived from decomposed bones, teeth, and crustaceans." *

After seeing and examining the above nodules from the crag and London clay, his attention was next drawn to the nodules in the green sand of Cambridge-shire, as likely to be found of equal value in agriculture. In a communication to the *Gardener's Chronicle* in 1848,† he says :—

" It occurred to me that possibly certain nodules of an anomalous character, abounding in the upper green sand in the neighbourhood of Cambridge, were in some respects allied to those in the crag, and would possibly be found to contain phosphate of lime. Upon directing the attention of Mr. Deck, of Cambridge, who is a practical chemist, engaged in making analyses for agricultural purposes, to these nodules, he readily detected in them the presence of earthy phosphates, in proportions varying from 57 to 61 per cent. Whether these various nodules, thus abounding in phosphate of lime, can be made available for agricultural purposes, must depend upon the possibility of their being collected at a cheaper rate than an equal quantity of bones can be."

He little thought, when he wrote this, to what extent, within a few years after, these nodules, or coprolites as they are still called in Cambridgeshire, would come to be used in agriculture, and the value that

---

* Proceed. Geol. Soc., 1843, p. 281 ; Rep. Brit. Assoc., 1845, sect., p. 51 ; Id., 1847, sect., p. 64 ; *Gard. Chron.*, 1851, p. 764.
† P. 180.

would be set upon them. The stratum of green sand in which they are found, "although never more than a foot thick, occurs near the surface over many square miles in the vicinity of Cambridge," and the land is now in places "honey-combed with pits" from which they are dug, finding employment for large numbers of labourers, and bringing in immense profits as well to the farmers as to the proprietors of the soil. The same operations are being carried on in some of the adjoining counties. Yet, great as was the benefit here conferred by Professor Henslow upon the farmer, no acknowledgment was ever made of his services. He never, indeed, looked for any himself. It was not his habit to work for reward. He was satisfied with having made a useful discovery, and it was no sooner made than he "at once gave it the widest circulation in the local papers, without reservation of any kind," in the same liberal spirit in which he always acted for the advantage of others rather than of himself.

It should not be omitted to state, in connection with this subject, that among the concretions found by him in the red crag at Felixstow, were some which proved to be the petro-tympanic bones of at least three species of large cetaceans, and additions to the list of crag fossils. These were described by Professor Owen in an appendix to Professor Henslow's paper in the "Proceedings of the Geological Society" above alluded to, and again afterwards in his own "History of British Fossil Mammals and Birds," where the several specimens are figured.

In 1845, he had an opportunity of bringing his scientific knowledge to bear practically upon a matter

of domestic economy as well as upon the more important concerns of agriculture.   This was in connection
with the potato disease, which first appeared that year,
and which, notwithstanding all that has been done and
suggested with the view of checking the evil, has prevailed more or less every year since.   " At a meeting
of the Hadleigh Farmers' Club on Friday, September
13th, the lamentable failure of the potato crop formed
the principal subject of discussion.   In the hope of
allaying the alarm which prevailed in the neighbourhood," he drew up a statement upon the subject, intended, in the first instance, for the farmers' use, in
which he gave general directions for the management
of the infected crop, and made suggestions respecting
the uses to which the decayed potatoes might be put.
A few weeks after, however, finding many applications
had been made for copies, he drew up a further statement in reference to this last point, embodying some
additional information, which he circulated more widely.
Some persons had thrown away large quantities of potatoes, which, when only partially decayed, he showed
to be an unnecessary waste.

" It seems," he says, " to be a providential arrangement, that as yet the really nutritive portion of the
potato is very little injured; even in those tubers
which have become partially decayed, and appear to be
wholly unfit for food.   The nutritious portion of the
potato consists of delicate white grains of starch-like
matter, which are enclosed in little cells.   When the
cells are broken the grains fall out, and, collecting
together, form a beautifully-white flour.   It is very easy
to separate this flour from the rest of the substance of

the potato; and if a few persons in different villages
would undertake to make the method generally known
among the poor, a vast amount of wholesome food may
yet be secured to them, which otherwise they will
suffer to perish. From an experiment that has been
tried, it appears that where 12 lbs. of flour can be ex-
tracted from a bushel of sound potatoes, 8 lbs. can be
procured from such as have become so far decayed as
to be useless as an article of food."

He then described exactly the process by which
this flour might be obtained, and the best method of
storing it up for future use; how, also, it might be
used for bread, if mixed with wheat flour in certain
proportions; or yet more for puddings, or still better,
"under the forms to which arrowroot may be applied;"
adding, " it is well understood that a very high per-
centage of what is sold in the shops under the name
of arrowroot, is nothing more than this very flour of
potatoes. It is also passed off in London under about
a dozen different names, as an important and nutritious article of diet."

Not content with issuing these directions, he gave
a lecture upon the subject in a few places, sometimes
publicly, sometimes in private families, having furnished himself with the necessary apparatus, and
going through the process in all its details, in order
to let his hearers see more exactly the way in which it
was to be carried on. The trouble he took on this
occasion, and the interest he excited on the question,
was not without its fruits. Many persons acted upon
his suggestions; one farmer, the Vice-President of
the Hadleigh Farmers' Club, "employed women to

grate about thirty sacks of bad potatoes," in order to obtain the flour; "at the Cosford Union-house about forty sacks of potatoes, too bad for use," underwent the same process; while, in his own parish, some of the poor were induced to turn to account in this way the smaller supplies which their gardens yielded of the vegetable on which they had always depended, but which failed them that season in so unexpected a manner.

Though so constantly engaged in different ways, independently of his ministerial duties; ever busied with the various schemes he set on foot for the improvement of his parish, and for years having to contend with the opposition of his farmers to the introduction of the allotment system; then, when that system was at length fairly established, exercising the closest supervision over the allotments themselves, encouraging the men in their work and stimulating them to industry; instituting experiments on manures, and looking closely into the relative values of crops raised under different circumstances; attending further to the diseases of corn, besides making observations for his own amusement, and collecting specimens in every department of Natural History; in addition to all this, he found time to interest himself in many of the scientific questions of the day, and was always ready to take up, in concert with others, any proposed investigation connected with his particular pursuits, and likely to lead either to useful results, or to the establishment of scientific truths about which there had been previously much doubt and difference of opinion.

Thus, in 1840, he was one of a Committee appointed by the British Association, to draw up a " Report on the Preservation of Animal and Vegetable Substances," and he himself undertook to conduct the necessary experiments for this purpose. These experiments were carried on for several years, and put him to trouble and inconvenience, independently of the time they took up, a part of the Botanical Museum at Cambridge being set aside for the reception of some 200 jars and bottles, containing the necessary chemical mixtures, and the various specimens to be operated upon, which last had to be inspected at intervals, and the results noted down. Yet he took great interest in this inquiry, from its especial use in assisting him to preserve many of the botanical specimens in the above museum, such as fungi, &c., which could not so well be exhibited in the dry state.

Another investigation in which he took an active part respected the length of time for which seeds, when preserved in different ways, retain their vitality. This question, of great importance to gardeners and agriculturists, had reference also to stories in circulation of seeds, obtained under unusual circumstances, having germinated, after having lain dormant, as was supposed, for a very long term of years. With a view to its solution, experiments were instituted by another Committee appointed by the British Association, of which Professor Henslow was one. The inquiry continued from 1841 to 1857 inclusive, and the results were embodied in sixteen successive Reports, published in the yearly volumes of the Transactions of the

Association. Some of the seeds experimented on were sown at Oxford in the Botanic Garden, under the superintendence of Dr. Daubeny and Mr. Baxter; others at Chiswick; and others at Hitcham, in Professor Henslow's own garden, and watched by himself.

So far as these experiments warrant the conclusion, the seeds which retain their vitality the longest belong to the *Leguminosæ*; one of these, a species of *Colutea*, vegetating after having been kept forty-three years, which is the utmost limit attained in any case. Next in point of duration of vitality are the *Malvaceæ* and *Tiliaceæ*, in both of which families seeds vegetated of the age of twenty-seven years. In only two or three instances in other families, seeds vegetated which had been kept twenty-one years; and in by far the greater number of instances, they failed to grow after having been kept more than eight years.

These results seem to throw discredit upon many generally-received statements as to the extreme length of time for which seeds in some cases can retain their vitality, especially the mummy-wheat, as it is called, found in the catacombs of Egypt. Professor Henslow never was a believer in any of the reported cases of the kind last mentioned. In certain instances in which such seeds had been submitted to his examination, he detected sources of fallacy, arising from the insufficient care that had been taken to preserve them from all mixture with other seed;* and he often insisted upon the extreme caution that was required to conduct experiments of this nature, a caution seldom

* See Rep. Brit. Assoc. for 1860 (Trans. Sects.), p. 110.

exercised to the necessary extent by mere practical men, who are not accustomed to that scientific accuracy which can alone insure trustworthy results.

In addition to seeds collected from the parent plants in the usual way, and sown by the above committee in 1841, the year in which it commenced its labours, many seeds of old dates taken from herbaria formed seventy and one hundred and fifty years back, others found in Egyptian catacombs and obtained from the British Museum, consisting of wheat, barley, and lentils, a few seeds of maize from Peruvian graves, and other seeds of old date, were all sown, principally in the Oxford Botanic Garden, with only negative re-sults,—" no vegetation taking place in any of the cases." If the experiment, as regards seeds of such antiquity, failed in the hands of a committee of scien-tific men, it is not very likely to have succeeded in any other. Further experiments are still wanting to determine the number of years for which seeds may retain their vitality, when buried at a great depth in the earth, and in nature's own keeping; but this is under different conditions from any of those to which they are ordinarily subjected when preserved by man, and the inquiry is not one that could be easily prose-cuted.

For the same reason that he disbelieved stories of mummy-wheat having germinated, he was slow to accept statements respecting the transmutation of species of plants which had previously been almost universally considered distinct. That is to say, he thought experiments in reference to that question needed to be repeated over and over again by diffe-

rent observers, under more rigid conditions than would
seem to have been secured, and the growth of the
plants more vigilantly watched than had been done,
in some reported cases of this nature, in order to
arrive at any certain conclusions. But he was nowise
prejudiced against the possibility of such changes
within certain limits. Judging from the present state
of our knowledge respecting the wide range of variation which undoubtedly takes place in some species
when cultivated, or growing naturally, under varying
circumstances of soil, light, air, elevation, &c., he
thought that we were far from having arrived at the
whole truth on this subject. At the meeting of the
British Association at Cheltenham in 1856, Professor
Buckman detailed the results of experiments carried
on by himself, in the gardens of the Agricultural College at Cirencester, from which he was led to infer
that several allied species of grasses, as well as certain
allied species of plants in other families, passed gradually one into the other, when under cultivation in open
borders. These experiments, which led to a long discussion at the time, Professor Henslow did not think
conclusive. Nevertheless, he communicated to the
Natural History Section at the same meeting, some
which he himself had made in reference to the
asserted transformation of *Ægilops* into wheat. He
said that "he had so far succeeded in changing
the character of *Ægilops squarrosa,* as to lead him to
conclude that M. Fabre's original statement, that
Æ. ovata was the origin of the domestic wheat (*Triticum sativum*), was not altogether without foundation.
He exhibited specimens, in which the form of *Ægilops*

squarrosa had undergone considerable change; but he had not succeeded in obtaining the characters of *Triticum sativum.*" On the same occasion, he produced other species of plants, reputed as distinct by some botanists, which he had undoubtedly proved by cultivation to be mere varieties; he also referred to instances of species of *Rosa, Primula,* and *Anagallis,* which he had found in the wild state decidedly passing one into the other.

For some years Professor Henslow set apart a portion of his garden at Hitcham, for such inquiries as the above, which he considered of increasing importance each year, from the excessive degree to which species had become multiplied, many resting upon very slight and insufficient characters. It was not till 1859, three years after the Cheltenham meeting of the British Association, that Mr. Darwin's book appeared "On the Origin of Species." It may be supposed with what interest he read this work, embracing the whole question of species in its most comprehensive aspect, and coming from a naturalist of such high reputation, and who had enjoyed rare opportunities of making observations in so many different parts of the world, independently of experiments closely carried on at home, connected with the breeding of animals and the cultivation of plants. He has nowhere formally recorded any detailed opinion of this work, though what he thought of it may be gathered, in a general way, from a letter of his to the editor of *Macmillan's Magazine,* in reply to some who had placed him in the ranks of those who supported

Darwin's views.*   He says, " The manner in which
my name is noticed in a review of Mr. Darwin's work
in your number for December (1860), is liable to lead
to a misapprehension of my view of Mr. Darwin's
' Theory on the Origin of Species.'   Though I have
always expressed the greatest respect for my friend's
opinions, I have told himself that I cannot assent to
his speculations without seeing stronger proofs than
he has yet produced."   He then alludes to a letter of
mine to himself, in which I had stated that I could
imagine that many of the smaller groups, both of animals and plants, might at some remote period have
had a common parentage, though I thought that there
was no satisfactory evidence to show that such had
been the case with the higher groups ; I added that I
did not, with some, say that the whole of Darwin's
theory *could* not be true, but that it was very far from
proved, and I doubted its ever being possible to prove
it.   To all this he said that it " very nearly expressed
the views he at that time entertained in regard to
Darwin's theory, or rather hypothesis, as he should
prefer calling it."   In this instance, as in so many
other instances, he showed his philosophic caution.
He declined going a step further than he could well
see his way ; at the same time that he would not take
upon himself to say that such and such things could
not be, where there were no sufficient data for establishing the truth one way or the other.   In previous
letters to myself, he had told me he thought there
were in Darwin's book too many suppositions, too

* *Macmill. Mag.*, vol. iii., p. 336.

many things assumed, which might or might not be true. Moreover, the further the question was followed up towards its source, the more it was beset with difficulties which were never likely to be solved. In fact, he said, he considered an inquiry into the origin of species about as hopeless as an inquiry into the origin of evil.

With reference to the religious aspect of this question, he stated, during his last illness, that he thought it an objection to the hypothesis of all animal and vegetable forms having been evolved in succession, through countless ages, from one primitive germ, or from a few such germs, that it did not allow for the interposition of the Almighty. "God," he observed, " did not set the creation going like a clock, wound-up to go by itself, but from time to time interposes and directs things as he sees fit." Yet he had always defended Darwin from some of his opponents, whom he considered as having shown him ungenerous treatment in fastening upon him opinions of an infidel or irreligious tendency, which those who are acquainted with his real sentiments on matters of religion know to be utterly without foundation. In the letter above alluded to, he mentions " a most eminent naturalist, who is himself opposed to, and has written against, its conclusions (Darwin's book), but who considers it ought not to be attacked with flippant denunciation, as though it were unworthy consideration;" adding himself: "If it be faulty in its general conclusions, it is surely a stumble in the right direction, and not to be refuted by arguments which no naturalist will

allow to be really adverse to the speculations it
contains."

Another subject in which he deeply interested himself, up to the time of his last illness, was that of the
flint implements found in the drift, and the question of
the antiquity of the human race as affected by the
evidence which those relics afford.   Tertiary geology,
and the recent changes of the earth's surface, had
always great attractions for him.   To the former, his
attention had been much drawn from the period of his
first visit to Felixstow, in 1842, a place he resorted to
with his family on several subsequent occasions during
the summer months, and where the red crag, in which
he first noticed the phosphate nodules, is so conspicu-
ously developed.   With respect to the flint implements, his opinions underwent, from time to time, some
modifications, and what his latest views were on the
subject are not known for certain.   He twice visited
the brick-pits at Hoxne, where these implements were
first found, and closely investigated the spot, and the
circumstances under which the celts had been met with.
His first opinion, after his return from that place in
November, 1859, was that the supposed antiquity of
the celts was a complete mistake.   He had no doubt of
their being genuine—*i. e.* of human origin—but he
believed them to have "all come from the vegetable
soil *above* the undisturbed beds of brick earth containing mammalian remains, fresh water shells, &c.," and
that, consequently, they were not coeval with these last,
as had been supposed.*   Afterwards, in another com-

* *Athenæum*, 1859, pp. 668 and 853.

munication to the *Athenæum,* dated Feb. 3, 1860,* in
consequence of further information he had received, he
stated that he had no doubt remaining on his mind
that these celts did "occur in undisturbed drift, but it
seemed to him likely that this drift may be a much
more recent deposit than some geologists are disposed
to believe."

Subsequent inquiries led him still further to modify
this opinion, as well as to regret the publication of
his earlier letters in the *Athenæum,* otherwise than as
serving to inculcate caution in the inquiry.   He continued his attention to the subject, comparing the researches of others with his own conclusions, and holding himself ready, if truth required, to abandon any
opinions which it might be shown he had too hastily
taken up.   With a view to obtaining further evidence,
in the autumn of 1860, he went to France to examine
the celebrated gravel-pits at Amiens and Abbeville,
where the same flint hatchets had been found in large
quantities ; likewise associated, as was stated, with the
bones of extinct quadrupeds.   The result of his inquiries in that locality, the geology of which he
thoroughly explored, at the same time "studying the
museums and collections in the neighbourhood," he
published in two other letters in the *Athenæum,*† in
which he still continued to give his judgment that the
beds in which these remains of primæval art occur were
not of that great antiquity which some have imagined ;
"neither did he consider that the bones of the extinct
animals found associated with the hatchets must *of ne-*

---

* No. 1685, p. 206.
† *Athenæum,* 1860, pp. 516 and 592.

cessity be supposed to have belonged to individuals contemporary with the workmen by whom the hatchets were wrought." His mind, however, afterwards underwent further changes, though to what extent, as already observed, is not exactly known. From a published extract of a letter of his, written in February, 1861,* it is evident he did not, up to that time, admit all the conclusions to which some other inquirers had arrived. He says :—

" I am intending next week to deliver a lecture at Ipswich on the pre-Celtic celts, which are confounding all our geological notions, and turning the world upside down, in regard to received chronologies. I strenuously advise *caution*, and repudiate some of the inferences which have been deduced from these remarkable discoveries. We shall hear of more of them. A new locality has just turned up at Herne Bay."

The above lecture was delivered the same month in which he wrote the letter alluded to, and a full report of it appeared in the local papers at the time. It possesses an interest independent of the subject treated of, as being the last lecture he ever gave.

He is said to have been preparing to lay his final conclusions on the celts question before the Cambridge Philosophical Society at the time of his being taken ill; and Dr. Hooker " believes that he had, at last, convinced himself that these implements belong to a period long antecedent to that usually attributed to man's existence on the earth, though by no means so distant as some geologists suppose."

From the beginning, he took up this inquiry " with

* *New Edinb. Phil. Journ.*, 1861, vol. xiv., p. 171.

a mind divested of all prejudice " one way or the other. He was quite prepared to accept any evidence based upon facts, and the inferences that necessarily followed such evidence, though militating against generally-received views ; as he had been " ready, in early life, to admit a conclusion which a practical acquaintance with geological evidence then satisfied (him) must be the truth, in regard to the inconceivable antiquity of the earth." He was not one of those who fear a collision between Scripture and science. His opinions on this subject, as expressed in one of his letters in the *Athenæum,* above alluded to, deserve to be recorded in his own words :—

" I dare to assert that I yield to no man in firm belief that ' all Scripture is given by inspiration ; ' but, then, given only for the purposes specified, viz. ' for doctrine, for reproof, for correction, for instruction in righteousness.' I am equally satisfied that proofs have been established, by arguments conclusive to all who have learnt to appreciate the evidence, that the inspired writers were often left to convey their lessons in their own words, intelligible to those whom they addressed, and in accordance with their own imperfect or erroneous views of nature. I, therefore, feel neither hopes nor fears in regard to whatever conclusions a thorough investigation of the question before us may elicit. Surprise, and great surprise, I presume we may lawfully feel, if it shall be shown that the views hitherto entertained by geologists and historians in regard to the very recent origin of the human race have been chronologically defective by many thousand years. It is the

L

misfortune of our day that so large a majority of all classes of society have not been educated sufficiently, in regard to the works of nature, for *believing it possible* that such questions as are now at issue can be satisfactorily set at rest by scientific induction. Yet, they who can look back a few years will remember how the same pulpits, then rebuking and maligning the conclusions at which geologists had arrived, are now content to accept them as evidence of a Wisdom, Power, and Goodness, beyond any that former ignorance could ascribe to the works of that First Great Cause, 'which spake the word, and they were made; commanded, and they were created.'"

In another passage, at the end of the same letter, after some remarks and speculations as to the period when, and the circumstances under which, the gravel in the pits at Amiens and Abbeville might have been deposited, he says :—

"If such a supposition" (the one he had made) "will not meet the facts, and a different conclusion shall be made palpable, we have only to be thankful that knowledge will have been increased. It is impossible to ignore the Bible in these investigations; but we have a right to expect that every link in the chain of evidence forged to controvert its *seeming* testimony should be most carefully scrutinized before its value as a holdfast can be admitted. We have cast off old prejudices erroneously deduced from the letter of the Scriptures, in regard to the age of the earth; but we cannot cast off our received opinions in regard to the time which man has inhabited the earth, without first feeling assured that these hatchet-bearing gravels must be several thousand years older than the pyramids of Egypt."

If, as appears to have been the case, he *was* led, in the end, to cast off his former opinion respecting the comparatively-recent origin of man, we feel sure that he never would have given utterance to any different sentiments from those here expressed, in respect of the reverence due to Scripture, as well as in respect of the true spirit in which alone all philosophical inquiries of this nature can be successfully carried on and brought to an issue.

CHAPTER VIII.

MANY as were the different subjects connected with science and the arts to which Professor Henslow turned his attention, there is yet another subject, hitherto unmentioned, that of Roman British Antiquities, in which, at one period of his life, he took great interest. He is believed to have acquired a taste for these researches from the circumstance of having been present at the opening of the celebrated Bartlow Hills, situate on the borders of Cambridgeshire and Essex. These hills consist " of a line of four greater barrows, and of a line of three smaller barrows," distant from the former between seventy and eighty feet. The smaller barrows had been opened, in January, 1832, by permission of Viscount Maynard, the proprietor, the excavations being under the direction of Mr. Gage, afterwards Mr. Gage Rokewode; and it was the opening of the first of the greater barrows, on April 21st, 1835, at which Professor Henslow was first present. He was then staying at Audley End, together with the present Master of Trinity, Dr. Whewell, Professor Sedgwick,

the Rev. J. Lodge, Librarian of the University, and
others; and a large party was made, including Lords
Maynard and Braybrooke, with their respective families, to witness the opening; the results of which, with
the contents of the chamber within the barrow, having
been fully described in the "Archæologia"* by Mr.
Gage, who, as before, conducted the excavations, there
is no need to allude to further here. The second of the
large barrows was opened in April, 1838, and the two
remaining large ones in April, 1840. There was nearly
the same party assembled on these occasions as on the
occasion of opening the first of the large barrows, and
the fruits of the researches have been similarly described by Mr. Gage.† Professor Henslow was, also,
again present both times, and thus had his interest in
antiquities kept up, until the time came when the
knowledge he had acquired of the subject was brought
to bear upon the examination of remains obtained in
other quarters.

These remains came, in the first instance, from Colchester; and consisted of fragments of Roman pottery,
which are occasionally dug up in great plenty in some
of the fields adjoining that town, well known to have
been a Roman station. His attention was drawn to
the circumstance by the late Mr. John Brown of Stan-
way, in company with whom he visited the spot, and
caused a quantity of the fragments to be brought to
Hitcham, from which he had the skill and patience to
reconstruct several amphoræ (earthen vases of very
large size, some three feet high or more), so completely

* *Archæol.*, vol. xxvi., p. 300.
† *Archæol.*, vol. xxviii., p. 1, and vol. xxix., p. 1.

and so neatly as to leave but few traces of their former broken condition. The best specimens of these reconstructed amphoræ he gave to the Colchester Museum; four that he kept for himself were to be seen ornamenting his hall, along with other curiosities to be hereafter spoken of.

But the chief occasion, subsequent to the opening of the Bartlow barrows, on which he had an opportunity of indulging in antiquarian researches, was the discovery of Roman remains in barrows at Eastlow, in Suffolk, in the years 1843 and 1844. "Eastlow, or Eastlow Hill, is the name given to a large barrow in the parish of Rougham," near Bury St. Edmunds. "The Saxon word 'Low' signifies a barrow. Three other barrows of small dimensions range in a continuous line with the large one, trending from it in a south-west direction." The contents of the most northerly of the three small barrows were accidentally discovered by some labourers in July, 1843, whilst engaged in removing the superincumbent earth for agricultural purposes. They had been found within a small brick chamber, and consisted of "a large iron lamp, with a short handle, and a very large and thick wide-mouthed square jar or urn of green glass, containing a large quantity of burnt human bones."

These remains being shown to Professor Henslow attracted his curiosity, and he was present himself at the opening of the small barrow, next to the one above spoken of, on the 15th September, in the same year. This proved richer in antiquities than the former one. There was the same brick chamber or vault at the depth of about six feet below the middle of the barrow, "which chamber, from its containing burnt human bones, forms the description of tomb called *bustum*." Within the *bustum* were discovered a handsome urn, or *ossorium*, of pale bluish-green glass, "which had fallen to pieces, and the fragments lay in a confused heap with the bones in the north corner of the chamber"; a lachrymatory, or vessel for perfume, composed of dull green glass, found "lying on the top of the mass of bones and fragments of the broken *ossorium*, and which had evidently been dropped into the urn after the bones were placed there; a coin, in a state of complete corrosion, and not to be determined, "found among the burnt bones"; two small jars, one plain, the other ornamented; a large spherical pitcher or jug of coarse yellow pottery; another jug, very similar to the last, but much smaller; two *paterae* of dark red ware, one a trifle smaller than the other, containing fragments of bone; two *simpula* of similar ware with the *paterae*, both found resting on their sides, with their bottoms against the south-west wall; "an iron lamp suspended from the extremity of a twisted iron rod driven horizontally into the south-west wall"; and, lastly, two iron rods, three and a half inches long, and slightly curved, supposed to have been "the handles of a small wooden chest which had gone to decay, but some traces of which were to be seen, in the form of carbonaceous matter, lying in the east corner."

All these vessels were most accurately described and figured by Professor Henslow, at the request of Philip Bennet, Esq., of Rougham, on whose estate they were discovered, and who allowed them to be exhibited at a bazaar for the benefit of the Suffolk General Hospital.

Professor Henslow's tract was published and sold for
the benefit of the same institution. It is accompanied by a lithograph, in which is given an exact
representation of the vessels and other remains, in the
same relative position in which they appeared in the
vault when the barrow was first opened. He has also
given separate figures of the *ossorium*, as restored by
his own hand, and of the lachrymatory. The restora-
tion of the former required great care, not so much on
account of the number of the fragments, some of which
"had entirely disappeared," as on account of those
which were found being "in a more or less advanced
state of disintegration." On this subject he has made
some interesting remarks. He says:—

"The manner in which the glass disintegrates is by
peeling off in small filmy scales, thinner than the
finest gold-leaf, or even than a soap-bubble; and a
puff of the breath scatters them through the air in
innumerable spangles, glittering with the colours of
the rainbow. As these scales fortunately peel off parallel
to the outer and inner surfaces only, and not along the
fractured edges, each fragment retains its original out-
line, and merely diminishes in thickness, so that they
could be restored with precision to their proper places,
though it was a work of some little labour to fix them,
since many were not thicker than the glass in a common
Florence flask. Before the urn fell to pieces its inside
had become partially encrusted with carbonate of lime,
which had crystallized in concretionary lumps, running
into each other so as to present a mammillated surface
internally, and a smooth, shining surface where the concretions had been in contact with the glass. Little

spherical, concretionary masses of carbonate of lime were also intermixed with the bones and dust in the general heap."

The *ossorium* had "two broad reeded handles, and an eared mouth." When restored, it stood eleven inches high; the neck was four inches, and the diameter of the eared mouth five inches, with the opening three inches in diameter; and it had a foot four inches in diameter, and an inch deep. The body was nearly spherical, and more than nine inches in diameter.

With respect to the lachrymatory, Professor Henslow was of opinion that these vessels, which are so often found associated with such remains, "were not tear-vessels, as is almost universally believed, but vessels for balms and balsams." It may be readily imagined "that perfumes, scattered over the remains of the deceased, became mingled with the tears of weeping relatives who were reclining over them, without our supposing a lachrymal vessel to have been handed about to collect these tears, in order to mix them with the perfume."

The last of the small barrows at Eastlow was opened the week after the one just described, but nothing was found in it except two broken earthenware vases, "each containing a few bones in an advanced state of decay," and a few other fragments of pottery. "Excavations were made in different directions, but no signs of any chamber were discovered."

On the 4th of July, the following year (1844), the large barrow, Eastlow Hill itself, was opened, and Professor Henslow was again present. The result was very different from what had been anticipated. "Instead

L 3

of urns and vases, *paterœ* and *simpula*, lamps and lachrymatories," such as had been found the previous year,
"the only contents of a large chamber of masonry
proved to be a leaden coffin, enclosing a skeleton."
This tumulus he has described in full detail in a second
tract, illustrated by woodcuts, being a reprint from the
Bury Post, in which the account was first published.
"The object of peculiar interest, to (him), was the well-
built chamber of masonry," which he supposed to be
"a solitary *existing* example of the manner in which
the Romans tiled their houses." Whether right or
wrong in this conjecture, the rarity, at least, of such a
structure gives great value to the minute record he has
left of the exact way in which the whole chamber appeared to have been built. An additional interest was
derived to the chamber from its arched vaulting, "a
mode of construction, of which," he believed, "there
are very few examples among us which can positively
be assigned to the Romans; so few, indeed, that at one
time it was imagined that they were not well acquainted
with the principle of the arch." In this instance, he
thought the arch had probably been turned over the
chamber after the coffin had been laid in it, when the
walls had been raised to the height of two feet.

In connection with all the circumstances of this
tomb, he said in his tract he should "permit (his) pen
to wander a little into the regions of imagination"; and
the reflections he has indulged in are such as must have
made his original letter in the *Bury Post* amusing as
well as instructive to general readers. He had, indeed,
the rare art, in his writings, as well as in his oral addresses, when intended for the people, of diffusing an

interest over subjects in themselves dry and unattractive.
At the same time, he would avail himself of every
opportunity, as he has done here, of inculcating any
moral or religious lessons that seemed to offer themselves. He has first sketched what he conceives may
have been the position in life of the Roman, whose remains had been lying mouldering in this narrow vault
" for 1500 years, or thereabouts."

" We feel confident that he must have been a person
who, in his brief day, had been eminent in some way or
other—for his wealth, or his rank, his valour, or his
position in the social system. No one, of little estimation in the eyes of his fellow men, would have been
buried in the style of this Roman—in a leaden coffin,
within a solidly-built vault, and with a monumental
mound of earth piled over it, which needed the united
efforts of a numerous company for its erection. I think
we shall not be wandering very far from the truth, in
supposing this person to have been lord of that neighbouring villa, whose foundations we detected last year
in a field at a short distance from these tumuli. He
was possibly the very last who died in occupation of it,
before the Roman legions were finally recalled from
enervated Britain, in the year of our Lord 426. I argue
thus in favour of the late period at which this tumulus
was erected. The Romans in the earlier periods of the
Empire burnt their dead, almost universally."

He then refers to the other tumuli at Rougham, as
affording "examples of this latter custom, with the usual
accompaniments of those vessels in which the offerings
to the manes of the deceased had been conveyed to the
bustum, and deposited, with the burning lamp, to cheer

them on their way 'to that bourne from whence (as they supposed) no traveller was ever to return' to the enjoyment of light and life, in a resurrection of the flesh." He contrasts these examples with the appearances presented by this later tumulus; alluding to the hopes, better than those entertained by heathens, which Christianity inspires; yet, in conclusion, expressing a regret that we should so often have to witness, even at the present day, instances of a superstition almost as debasing as any that prevailed in the heathen world.

" No one (he says) rejoicing in our happier prospect, can look upon those relics from the smaller barrows, preserved at the Hall at Rougham, without feeling them to be a record testifying to the general belief of mankind in the immortality of the soul. But in the arrangements within this larger and later tumulus, perhaps we have some trace of the already-spreading influence of a still better creed. During the 400 years that the Romans held this country in subjection, the Gospel had been gradually leavening the corrupting mass of heathen superstition. Better conceptions of what is life, and what is death, were becoming interwoven with the current opinions of the world, and they were inspiring even heathens with a contempt for practices which could profit nothing to departed souls. The simpler mode of sepulture adopted for this Roman may have had some connection with that mighty revolution which was then taking place in the world of mind. The Christians were everywhere abandoning the practice of burning the dead; and, though their faith may not have reached the heart of this Roman, yet his head may have assented to better notions than those which had

persuaded his predecessors at Rougham to feed ghosts
with oil and wine, milk and blood, and other substantial
creations, suited only to the sustenance of a bodily existence.   For, where are those funeral rites which we
found had been so carefully attended to in the other
cases ?   The funeral pyre no longer blazes.   The lamp
is no longer considered of any importance.   No offerings are placed within the vault.   All that could be
found within the tomb indicative of heathen superstition was the pass-money (an *obolus*) in the mouth of the
entombed.   Charon had been propitiated.   I have not
yet been able to distinguish any legend on this coin,
which is nearly as much corroded as the one found last
year.   There was a little chamber outside the vault, in
which glass vessels had been deposited, but unfortunately these had crumbled to powder, and there was no
relic of any kind to show what they had contained.
If that rusty *obolus* had been missing, we might have
felt half persuaded to believe this Roman had embraced
the Cross.   The superstitions of those days, and of
later days, and, alas ! of these days also, are strange
things to look upon.   Indeed, we have no need to tax
our imaginations for what the false fancies of ignorant
and unenlightened minds may formerly have tempted
men to put their trust in.   I allude to none of the
vanities of will-worship; but it seems that even the record in the Acts, concerning those dealers in curious
arts who burnt their books and repented, is a lesson
lost upon many of us now-a-days; and we still hear of
hundreds consulting some 'wise man,' or 'wise woman'
(wise, indeed, in their generation), as confidently as this
heathen ever trusted an Aruspex or an Augur.   Truly,

a thousand years in these matters have passed but as
one day!"

After indulging in the above speculations, he proceeds
to describe the vault more in detail (the chief features
of which have been already noticed), mixed up here and
there, as usual, with an amusing observation, and then
adds some remarks upon the way in which the leaden
coffin or chest had been made, as also upon the condition of the skeleton found in it.   It is singular, that
attached to this last, he should have found something
to attract the attention of his botanical eye, whilst
busied in antiquarian research.

"There were root-like fibres projecting from the bones
(of the skeleton), of the legs more especially, which
gave them a strange and shaggy appearance.   This
proves to be a mass of a peculiar kind of fungus, called
Rhizomorpha,* and serves to illustrate the fact, that all
fungi are derived from the decomposing materials of
some previously-organized body, whether animal or
vegetable.   Here we have the substance of one of the
nobles of antiquity converted into materials forming one
of the very lowest of the fungi!"

---

* The genus *Rhizomorpha* is now sunk by Botanists, being
ascertained to be only an underground or subcortical form of
mycelium common to several species of fungus, where it is of
a fungal character at all ; but, like the Ergot, not capable of
further development, except under certain conditions.
   Many productions, however, which have been referred to
this spurious genus have no pretensions to a fungal origin, but
are mere conditions of the roots of Phænogams, like those
which occur in wooden pipes, or fill up drain tiles through
which there has been a constant flow of water.   Mr. Berkeley
thinks it probable that the substance found on the legs of the
skeleton spoken of above may have been of this latter character.

The skull of this Roman was forwarded to the Anatomical Museum at Cambridge, and the leaden coffin transferred to the Fitzwilliam Museum. At the conclusion of his tract, he mentions having collected some additional memoranda relating to urn burial, which, if he could find time to throw them into a presentable shape, he intended to offer in the form of a lecture to the inhabitants of the neighbourhood in which the above remains were found. I am ignorant whether this lecture was ever delivered. No one can read the two tracts just noticed without being satisfied that he was fully competent to undertake it, notwithstanding, with his accustomed modesty, he apologizes for his " very slight acquaintance with antiquities."

CHAPTER IX.

FROM the survey that has been now taken of Professor Henslow's labours, it will be seen how diversified were the subjects to which he devoted his time and consideration. The circumstances in which he was placed, with a large parish to attend to, naturally, in a great measure, determined the chief occupations of all the latter part of his life. But had he been independent of those circumstances, and had he restricted his researches to a narrower field, his abilities and modes of thought were such as would have secured the most valuable results for science in any department he might have taken up—results far beyond those to which he actually attained. Either in Botany or Geology, the two branches of Natural History to which he seemed to have the strongest leaning from within a short time after graduating at the University, he might have fairly taken the lead among the botanists and geologists of his day. What he did accomplish in these sciences is quite sufficient to show what he might have undertaken, if his attention had not been called off in other directions. There can be little question that he who, at the early age of twenty-six, and only two years after receiving his first lessons in geology, worked out with such complete success the details of a difficult country like that of Anglesea, would, if he had kept entirely to that study, have proved, as life advanced and knowledge and experience advanced with it, fully equal to disentangling the most complicated phenomena the earth's structure presents, and discussing the highest problems connected with the mighty changes that structure has, from age to age, undergone, so far, at least, as these phenomena can be understood and explained, and these problems solved, by man with his present limited capacities.

His mind was of that philosophical cast that fitted him to deal equally well with details and generalizations. There was not a more accurate observer. No one, either, was more assiduous in collecting facts, and no one knew better than himself how to make use of them. He never hastily took up any hypothesis; at times, perhaps, he was rather over-cautious than otherwise in this respect, but he gladly accepted a well-supported theory that would serve to bring isolated phenomena together, however it might run counter to all his own previous habits of thought.

But he had not the leisure and opportunity, if he had had the will, for exercising his faculties in the way above described to the full extent of his powers. His lot was differently cast. And, as it is, his character as a scientific man will rest not so much on what he discovered or worked out himself, as on his indefatigable energy and consummate tact in organizing methods of acquiring useful and scientific knowledge; in suggesting and directing the best ways for conducting particular inquiries; in stimulating and assisting others, who enjoyed opportunities denied to himself, and putting them on the right track for successfully prosecuting their respective researches. The benefits he thus conferred on science were felt both by individuals and by public institutions. It is stated by Mr. Darwin, in his communication inserted in a former part of this memoir, that he owes more than he can express to Professor Henslow for his kindness and assistance; not only for having got him the appointment as Naturalist to the *Beagle*, but for regularly corresponding with him, and guiding his efforts, during the five years of his voyage with Captain Fitzroy, and taking charge of the collections he sent home. In like manner, Mr. Lowe, in a private letter to myself, speaks of the great and valuable kindness he received from him, when resident in Madeira; and that, in addition to other acts of friendly assistance, he undertook to raise for him a hundred pounds among the members of the Cambridge Philosophical Society, in aid of his Travelling Bachelorship, and to enable him to carry on his scientific pursuits. And there are others who could give similar testimony.

It was his strong desire to see the intellectual standard raised in all classes of society, joined to the conviction that no branch of knowledge was better adapted than Natural History to quicken the energies and improve the minds of young persons, as also to humanize their affections and to lead them to higher views of the Divine Wisdom and Goodness, that made him take so much interest in schools and education, and endeavour to diffuse a taste for Natural History pursuits. The public have here reaped the advantage of his appointment to a parish like Hitcham, where the people were so thoroughly degraded in mind and habits of life when he first came among them. It brought him into closer acquaintance with all the evils resulting from the neglect of education; the deeply-rooted prejudices which have to be overcome before any improvement can be attempted or instruction given; the ignorance, that sometimes resists the best-directed efforts to do good; the superstitions, which check the growth of all right feelings upon religion; above all, the vicious pleasures which usurp the place of healthy recreations and rational amusements. It was his allotted task to organize plans for the amelioration of this state of things; and the example he has left of what can be done towards it, when the mind and will both give themselves to the work, will assist others in coping with the same evils in other places.

Also, by employing science as a means to this end, he has utilized it for purposes to which it had never, perhaps, been so successfully applied before. Lectures upon certain branches of science have, indeed, been often given in parishes where the inhabitants had already made some advances in education; and Natural History has been occasionally taught in schools of a higher grade. But Professor Henslow availed himself of the sciences for raising an utterly-neglected people from the sinks of vice and ignorance. By engaging their hearts, in the first instance, with a sight of the wonders and curiosities which nature reveals to those who will look into them, it was not difficult afterwards to pass on to the civilization of their manners and the cultivation of their minds, until many of the young became ardent followers of him, even in his own particular pursuits.

Yet we must not look alone to what he did in his parish, or to the success that attended his lessons on Botany in the village school, much as the latter, from their novelty, may have attracted attention. Rightly to estimate his worth as a public teacher, we must take a wider view of the field in which he laboured. We must regard the high scientific and educational value of his lectures at the University; his examinations, for the Natural Sciences Tripos; his examinations, also, at the London University. It is here that his great abilities as a teacher are most conspicuous, his masterly tact in selecting the best ways of inculcating scientific knowledge. He did more than anyone to raise the teaching of Natural History, in particular, to its proper value as a method, and dignity as a pursuit. If a man felt a strong desire to take up Botany, in the hope of attaining in time to a complete knowledge of the science, there was no surer way of proceeding than beginning by an attendance on Professor Henslow's Lectures, which, though, of course, elementary, were devoid of all superficiality; as there was no better proof of the proficiency he was making than being able to pass the Professor's examinations. Or if, without any intention of studying the science deeply, he wished to take it up as an agreeable occupation for his leisure hours, he would feel encouragement to do so, and find his interest in the subject continually increase, as he heard, from time to time, the earnest language in which the Professor would descant on the claims of Botany to some share of our attention, its close bearing upon other important branches of human knowledge, the advantages it possesses above so many of the ordinary pursuits of life—yet more, the enlarged views which not Botany alone, but Natural History in general, gives us of the works of the Creator.

It would be unnecessary to repeat here what has been already said respecting the influence which Professor Henslow exerted over the minds of the young men at Cambridge, during the period of his residence in the University. But there is reason to think that it may have worked as much for good in that place on the upper classes, though not so openly brought under the public notice, as his influence manifestly worked for good at Hitcham afterwards in the case of the lower classes. It may have deterred some young men from entering upon courses of life, which, from tempting to guilty excesses, too often end in folly and shame; or which, if attended by nothing that the world calls sinful, have, at least, a degrading effect upon their faculties, causing regret to others as they will surely one day cause regret to themselves, that time and talents, instead of being made to serve for their own improvement, or for the good of their fellow-creatures, should be wasted upon pursuits utterly unworthy of man's place in the scheme of creation.

It was, however, to men called to such professions as of necessity require a certain amount of Natural History knowledge that Professor Henslow's teaching was of the greatest value. Medical men, especially, owe him a lasting debt of gratitude for instituting examinations, which placed Botany upon an entirely different footing

from that on which it had formerly stood in their estimation. From the imperfect way in which Botany had
previously been taught as a branch of medical education, and the little good the students had got from it,
the latter had been led to regard it as a useless or even
mischievous interference with other courses of lectures
having a more practical bearing upon the profession.
The botanical examination was often little more than a
mere test of memory in enumerating hard technical
terms for abnormal and unusual organs, or answering
physiological questions by rote; or it required the
writing of useless histories of once useful drugs, or the
giving generic or specific names of pharmaceutical
plants learnt the day before for the occasion. Of what
character Professor Henslow's examinations were, as
contrasted with the above, may be fully seen from the
account Dr. Hooker has given of them in connection
with the London University, in a former part of this
memoir.

We should form but an imperfect estimate of Professor Henslow's taste and love for science, if we were
to leave out of our consideration his fondness for collecting in connection with the different pursuits and
studies he followed up. His zeal and diligence as a
collector was not less remarkable than his closeness as
an observer, and his abilities as a teacher. The extraordinary assemblage of things he had got together was
such as we may safely say had never been seen before in
any parsonage. Nothing was passed over that touched
in the slightest degree on science and the arts, or that
helped to illustrate the lectures he delivered in different places, and his *lecturets* at the horticultural shows

in his own garden. Mention has been already made of the collections he formed for the Botanical Museum at Cambridge, and the rich contributions he sent, from time to time, to Kew and the Museum at Ipswich. Yet quite independent of all this, the Rectory-house at Hitcham was as thoroughly stocked with specimens of all kinds as if everything he had been able to obtain was centred there. It was itself a museum. The hall, the passages, and almost every room in the house to the very attics, had their respective valuables and curiosities. To attempt to enumerate all the different articles would be quite hopeless and it would be unnecessary; yet without some detailed account of what he had accumulated, it is hardly possible to form any adequate idea of its extent or of its great variety. Natural History, of course, was first considered, and formed the staple of the collections, and every branch of that subject was equally attended to and represented. In the Animal kingdom might be noticed—numerous cases of stuffed birds, and a few of quadrupeds,—bird-skins in boxes,—drawers of nests and eggs,—a human skull and skulls of quadrupeds,—horns and tusks of some of the larger mammals,—elephants' teeth,—calculi;—a crocodile six feet in length,—alligators, snakes, and other reptiles, and various fish in spirits;—crabs and other crustaceans,—cases of centipedes, scorpions, and spiders,—numerous drawers and glazed cases of British and foreign insects;—shells, British and foreign,—a series of specimens of the pearl oyster, illustrating the formation of pearls;—starfish, corals, madrepores, and sponges. In the Vegetable kingdom were several large presses of dried plants,—extensive collections of seeds, pine cones, fruits, gums, and other vegetable products, some in cabinets, and others in bottles, besides a series of wax models;—specimens of various kinds of woods;—specimens of the sugar-cane, and other foreign reeds, along with portions of gigantic fossil reeds;—cases of insects producing galls, and insects injurious to timber;—specimens of the different varieties of wheat in the ear, and other kinds of corn, with other specimens infected by the different diseases to which corn is subject. Of Minerals there was a large collection, besides drawers of polished marbles, agates, and pebbles;—several boxes of models of crystals in wood;—an extensive geological collection of rocks and fossils, in illustration of the different British strata;—plaster casts of some fossils;—lavas, and other volcanic products;—geological sections, and models and sections in illustration of mines and mining.

Such were the collections in the Natural History department. Those relating to the Arts and Antiquities were scarcely less remarkable for number and variety. The former included a series of specimens illustrating the various arts of hat-making, weaving, painting, dyeing, printing, and candle-making;—specimens in illustration of the manufactures of glass and pottery,—others illustrating wire-working and other iron-works;—specimens of silks, cottons, and flax, in different states;—samples of paper;—bottles containing specimens of the different kinds of oils;—specimens of wood-carving, and of Tunbridge ware;—patterns of veneers,—enamels, cameos, and mosaics;—models of various kinds—including canoes, ships, can-

non on wooden carriages, and a model of the Plymouth
Breakwater. Among the Antiquities were to be seen
the large Roman amphoræ, in the entrance hall, before
spoken of, along with cinerary urns, vases of different
kinds, lamps, lachrymatories, pateræ, and other specimens of Roman pottery;—coins and medals;—querns
and celts;—ancient guns and matchlocks. There
were likewise collections in illustration of the habits
of savage nations : idols of various kinds ;—dresses of
the inhabitants of some of the islands in the Pacific
Ocean ;—necklaces, bracelets, shoes, grass hammocks,
paddles, war-clubs, bows and arrows, spears, shields,
and other implements of war or of domestic use from
different countries.

In addition to all this there were numerous miscellaneous articles of interest and curiosity,—trophies
from Waterloo, relics from the *Royal George*, and
such like;—maps, diagrams, drawings, and classified
woodcuts, of which he is said to have "had literally
thousands, mounted and instructively arranged ";—
orreries, microscopes, magic and phantasmagoria lan-
terns, gyroscopes, chromatropes, and various other
kinds of philosophical instruments beyond specifica-
tion;—one room on the ground-floor being entirely
given up to some of the larger and less delicate kinds
of chemical apparatus, associated with jars and vessels
of all descriptions, ready for use as wanted.

After giving the above somewhat lengthened statement of the contents of his museum at Hitcham, it
must be observed that all these things were not " the
miscellaneous hoardings of a mere collector," but
specimens and objects carefully put away, rather for

M

the benefit of others than of himself, and intended
to serve for purposes of instruction as already mentioned.   It is surprising to think that, extensive as his
collections were, there was scarce a specimen that was
not fully labelled, with some little particulars of its
history.   Nor was there, perhaps, any object of general
interest which was not brought out in its turn, and
explained to some of the attentive thousands who,
during the period of his Hitcham life, must, at one
time or another, have had the opportunity of hearing
his popular addresses on Natural History and other
branches of science.   His museum was the store-house
from whence he drew forth, as required, the materials for
imparting useful and entertaining knowledge,—knowledge given freely, and scattered broad-cast among
all classes indiscriminately who were disposed to
come to his intellectual feast.

It should be yet further mentioned that so great
was his liberality in giving specimens away, that in
general he reserved for himself only duplicates.   In
all cases he gave the best specimens to public museums
—Kew, Ipswich, and Cambridge being always first
considered,— or to persons who he thought would
make the best use of them.   In so doing he was
actuated by the purest love of doing good.   He was
totally devoid of vanity : he never cared to see his
name made public,—never asked for anything in return, or looked even for acknowledgment.   This was
the more remarkable, since he had all a collector's
love of specimens, and since every specimen cost him
thought, manual labour, and often much money.

But the room of all others in the Rectory which

afforded the most unmistakable signs of the habits and pursuits of its busy occupier was his Study. Independently of a library of books which covered the walls, it was so crowded with papers and tracts, diagrams, boxes, bottles, and aquaria, apparatus of various kinds, plants in course of drying, insects on the setting-board, specimens of natural history under examination, or wanted in illustration of some subject on which he was at work, that it required a careful step and a vigilant eye on the part of visitors to avoid injuring or displacing its multifarious contents.

In this Study he would sit and work the greater part of the morning—sometimes till late in the afternoon. He would then generally go into the village,— which from its extent and scattered population, and its lying a little distance from the Rectory, afforded in itself a moderate walk,—if there was nothing to call him in any other direction. He would look in at the school, or go round and inspect the allotments, and notice what improvement had taken place in them, or visit the sick and other poor in their houses, according to circumstances. In the evening, after dinner, he would take, as he quite needed, a little rest; he would then divide the remainder of the time between reading and arranging plants, or preparing and mounting specimens for the Ipswich Museum or for his own collections, or amuse himself with some similar manual occupation as the case might be, till midnight, or not unfrequently beyond it.

He was quick and ready at all mental work, and got through a great deal in the day. Asking him on one occasion what he had been doing all the morning, he replied that he had made a sermon for Sunday, written eleven letters, and set an examination paper for the London University. This was an amount of work which few persons perhaps could accomplish at a morning's sitting. It showed the strength of his faculties; the readiness with which he could command his ideas, and clothe them in the right language, within a very limited space of time.

To a man of less powerful mind it would have been almost sufficient employment to attend to the voluminous correspondence he kept up. From the versatility of his talents, and the multiplied number of subjects in which he interested himself, combined with the wide reputation he enjoyed, he was constantly receiving letters in reference to the plans he was carrying out or which he had suggested to others, or in which the writers were seeking further information from him about his lectures, or stating the results of experiments on manures and other agricultural operations, or asking his assistance and advice for different purposes.

It should be stated that, in addition to all other demands upon his time, he acted as a magistrate, and for many years was regular in his attendance at the board, though distant four miles from his own house. Latterly, from his various occupations, he ceased to attend so regularly; but his name continued to be on the commission to the period of his decease.

Besides belonging to the Linnæan, Geological, and Cambridge Philosophical Societies, he was a fellow of several provincial societies for the promotion of general science, or of Botany and Horticulture in particular.

The Society for the Diffusion of Useful Knowledge, the Ray Society, and the Palæontographical Society, all received more or less assistance from him. His connection with the British Association and the interest he took in its proceedings have been already in some measure alluded to.   He joined that body in the second year of its existence at the Oxford meeting in 1832; and when it met at Cambridge the following year he was appointed one of the local secretaries. Those who remember the latter gathering must remember also the active part he took on the occasion; the energetic endeavours he made to get all parties to work cordially together, and to have everything arranged—some of the arrangements being entirely his own—so as to ensure success.   After taking up his residence at Hitcham, from being so much engaged in ministerial and parish duties, he was not able to attend the meetings of the Association very often; but he was always ready to join in the work of any committees in which he could be of use as before mentioned.   He was four times President of the Natural History section, the last time at Oxford in 1860, when the animated discussion took place in that section on the Darwinian views about species.   A large audience was drawn together to hear it; and those who were present on the occasion speak of the admirable tact and judgment with which he regulated the discussion, showing complete impartiality, allowing every one fairly to state his opinions, but checking all irrelevant remarks, and trying to keep down as much as possible any acrimonious feelings that appeared to

mix themselves up with the arguments of the con-
tending parties.

He had an extensive acquaintance, quite independent
of his numerous correspondents just spoken of, many
of the latter being personally strangers to him.   Nor
was his acquaintance at all confined to the scientific
world, widely known as he was there.   Both in and out
of the University, in which he resided for so many
years, and equally in the upper as in the middle ranks
of life, he had a large circle of friends, who highly esteemed him, and who sought out his society, not more
for the sake of his general knowledge on matters of
science, which he so freely imparted to all, than for the
unblemished excellence of his character, and the kind-
liness of his disposition.   His frank and open-hearted
address, and his agreeable conversation, made him a
universal favourite and always welcome.   His family
motto was "Quod videris esto," and he not only acted
upon this rule himself, but gave others credit for acting
upon it also.   His universal love and charity made him
slow to perceive faults in those with whom he associated; he never looked for faults, and he was unwilling
to be guided by any reports he heard to the disadvantage of persons of whom he had always thought
well.   Desirous of living upon friendly terms with all,
he seemed to regard all men as worthy of his friendship, and opened himself to them as if the human
heart were everywhere the same, and as utterly incapable of deceit in others as it was in himself.   Of
course, he was occasionally taken in; and during his
last illness he mentioned one instance in particular, in

which he had been "sadly deceived by the supposed righteousness of a wicked man." He said the circumstance caused him great misery, but he prayed for the individual in the hopes he might become reconciled to God. Yet, from the same unsuspecting nature, in matters in which there were no principles involved,— nothing that came into collision with his love of truth, morality, or justice,—he was disposed to believe everything that was told him with the simple-mindedness of a child, so that I have heard some of his nearest relatives jokingly observe, that he was one of the most charitable men on earth, as " believing all things."

He never spoke otherwise than as he thought. Though what he said might not always fall in with the views of the company in which he happened to be, it was said in a conciliating tone and manner that forbad the taking of offence. He would not allow a breach of charity to arise out of the friendly intercourse he wished to keep up with those about him. Though a stanch advocate for truth, and earnest in his endeavours to remove error and instruct ignorance, he did not interfere with the opinions of others who thought differently from himself on questions open to discussion. He could mix freely with men of all parties, without reference to their religious or political views, not concealing his own, yet not allowing theirs to lessen his esteem and admiration for whatever there was of good or honest in them in other respects.

Another remarkable feature in him was, the evenness of his spirits and the evenness of his temper. He was always the same and always cheerful; nothing put him out or ruffled him. Whatever might be said or done by others, whatever occurred to him in other ways, was alike taken in good part. He derived his happiness from sources quite independent of the outward accidents of life, and no man seemed to enjoy life more, or to be more thoroughly contented with his position in the world. And his desire and endeavour was to make others as happy as himself. Affable and easy of access, though busied from morning to night with his various occupations, he was never too busy to attend to those who came to him, to listen to what they had to say, or to assist them in any way they needed as far as was in his power. Nor was all this merely the result of temperament and constitution. Though in some degree assisted by these, it arose from higher motives. It flowed naturally from his firm religious principles, which led him conscientiously to discharge all his duties, to be useful in his generation, and while bearing with the infirmities of those weaker than himself, to seek their good in every possible way.

CHAPTER X.

SICKNESS AND DEATH.

THAT Professor Henslow was habitually under the influence of religious principles, it has been endeavoured to show in a former part of this memoir, when speaking of the spiritual ministrations that devolved on him as a parish priest. But if there were any doubt about it, what he thought and felt on the subject of religion was clearly evinced during the severe illness that preceded his death.

It was probably to his "incessant mental and manual labour" that his illness was to be attributed. He was indefatigable in his pursuits, and most careful of his time. He seldom allowed himself relaxation. He not only worked hard the greater part of the day, but had always some little employment at hand to fill up the odd minutes before meals, and other such intervals. He had naturally a very strong constitution, and in his younger days was capable of enduring great fatigue. I remember to have heard him say, that when geologising in Anglesey, he had walked forty miles in the day with a bag of stones at his back, and was yet equal to

dancing at a ball in the evening afterwards.   But the
strongest constitution gradually gives way, when either
the mind or body is habitually overtasked.   For much
of the latter part of Professor Henslow's life his mind
seemed always on the stretch, while his out-of-door
exercise was hardly sufficient to preserve health under
such circumstances.   It was about five years before his
death that he first complained of an affection of the
chest, for which he took advice, and which obliged him
for a time to live by rule, and to avoid over-exertion.
It was attributed to bad digestion, "though he was
always abstemious and temperate in every respect."   By
proper care, the symptoms were greatly alleviated, and
his health was to a certain extent restored.   In March,
1861, however, the complaint, which had been evidently
lurking in the system ever since the first attack, returned in an alarmingly aggravated form.   He had
left Hitcham about the middle of that month, though
feeling far from well, to visit some friends in the south
of England, where he caught a violent cold, followed by
the same oppression on the chest he had felt before,
and a difficulty of breathing when he walked.   He returned home very ill on Saturday, the 23rd, and grew
so rapidly worse during the following week, that on
Thursday, the 28th, by the advice of Mr. Growse, his
medical attendant in the neighbourhood, his son-in-
law, Dr. Hooker, was summoned from London immediately, and urged to bring a physician down with him.
It was thought at that time that he would not live
many days, and might not live many hours.   The following day, Good Friday, Dr. Hooker arrived, with Dr.
Walshe, and on the same day Professor Henslow took

to his bed, which he never left afterwards till the period of his death. Dr. Walshe pronounced it to be a complicated case of bronchitis, congestion of the lungs and liver, and enlargement of the heart. He gave no hope of recovery, though he thought the fatal termination would not take place so soon as had been imagined. This proved correct. Professor Henslow lingered on for seven weeks from the day on which Dr. Walshe had been sent for, during which protracted period of decline his sufferings were occasionally very great, and for the first two or three weeks all but continuous. After that time, they abated for ten or twelve days, and then returned. The paroxysms occasioned by inability to breathe freely were, it is said by those who waited upon him, as painful to witness, as they were agonizing to himself to bear. They were such as would have worn out, in much less time, patients of a weaker constitution. But his great natural strength gave way very slowly, and resisted, beyond all the first expectations of his friends, the encroachment of his disease. It was, to use his own expression, " like granite wearing away from drops of water." He was, indeed, quite aware of his circumstances, and most thoroughly prepared for what was surely, though thus slowly, approaching.

One singular feature, indeed, in his illness, was not merely the magnanimity and composure with which he received the intelligence of the nature of his disease, but the lively interest he took in his own case as a physiologist. All this was the more remarkable when contrasted with the state of his feelings on the occasion of his illness five years before. He was then seriously

alarmed for himself.   It was the first time in his life
in which he had been affected by anything more than
slight casual indisposition, and fear as to the result of
that attack took firm hold of his mind.   He had,
moreover, always considered himself as having a consti-
tutional idiosyncracy that rendered him unable to bear
pain.   But now that he was called upon really to suffer,
all was changed.   No sooner was he told on Good
Friday that he could not live, than he evinced from
that moment an utter indifference to his fate.   He immediately rose superior to all further desire for life, all
fear of death, and all shrinking from what he had to go
through before death would release him.*   In the face
of inevitably increasing sufferings, he set himself to
watch the successive symptoms of approaching dissolution, all of which he desired should be communicated
to him by his medical attendants, with whom he discussed them as a philosopher, and without the most
distant reference to himself as being the subject of
them.   His ability to do this, in connection with so
strong a feeling of interest in the progress of his
disease, was such as few sufferers to the extent to which
he suffered have perhaps before exhibited.   It was not
merely the same moral courage which he had shown
through life.   It was not simply the fortitude with
which so many besides himself have met their end.   It

---

* The above circumstance seems strikingly in accordance
with a remark by Sir B. Brodie in his "Psychological Inquiries."   Speaking of the actual fear of death, he observes :—
"It seems to me that the Author of our existence, for the most
part, gives it to us when it is intended that we should live, and
takes it away from us when it is intended that we should die."
—P. 128.

had a higher source and a deeper foundation than either. It had rather the appearance of an almost superhuman effort to suppress and keep under all thought and sense of what he was enduring in order to give his undivided attention to the circumstances of his case. His love of knowledge was unquenched to the last, and there was a philosophic contempt both of pain and dying while gratifying it. He seemed even to regard his affliction as a means offered to him of acquiring further knowledge than he had yet attained to, equally as sent for a far higher purpose. It found employment for his mind —very different, indeed, from the light amusements to which many would turn to while away a weary hour in the intervals of pain—but falling in with the predominant tastes and pursuits of his whole life, and such as he had only this last opportunity of following up in the world.

Yet let it not be supposed, from what has been said above, that his was the death of a philosopher, and not at the same time the death of a Christian. Never was there a more glorious manifestation of the faith and love which Christianity inculcates. During his whole illness he was a model of patience and resignation to the Divine will. He prayed that not a murmur might escape his lips. He expressed the most sincere gratitude to the Almighty for His mercies to himself, and placed his entire trust in the Saviour, with absolute renunciation of all personal merit. He observed, —" What a blessed thing it is to be a Christian,—and a blessed thing for a Christian to die!" He said he had not before his eyes, to his utter astonishment, that fear of death which he thought he should have. He placed his soul in the hands of a righteous Creator. He had confessed his sins, and he desired nothing more in this world: he asked for no return of health, no miracle to be wrought on his behalf,—only an easy death. He could not see that his confidence was too high. He regarded it as the highest blessing to know and feel that all his sins were blotted out by the Saviour; to be able to look them in the face, and say they were forgiven him for Christ's sake. At times he would exclaim—"Lord pardon all mine iniquities; pardon all my sins:"—repeating the text afterwards —"If any man sin, we have an advocate with the Father, Jesus Christ the righteous; and he is the propitiation for our sins;" and,—"This is a faithful saying, and worthy of all acceptation, that Christ Jesus came into the world to save sinners," adding, "save even those that are lost." Again and again he declared on different occasions that he died in the faith of Jesus Christ, and in the hope of eternal life.

He derived great consolation from reflecting upon his useful and well-spent life, and the thought that he had not a single enemy that he knew of. He often talked with Dr. Hooker on this subject, and reconciled it with his renunciation of all personal merit. He agreed with the latter that to suppose we were to be so tormented on our deathbeds with our sins and shortcomings as to derive no comfort from our earnest attempts to do good, would involve a paradox, and be refusing the consolations which our Maker grants to us in this life.

On the other hand he sometimes said he felt that he might have done more, with the same opportunities;

though after-reflection led him to remark that it was of no use fancying what can never be, and that it was better to employ well the time yet afforded him. He also incessantly feared lest he might be deriving too much comfort from reflecting upon his past life. He said that were there a possibility of recovery it might be very dangerous and lead to presumption or worse, and that as it was he might in this matter be over-confident. On its being remarked to him that it was scarcely possible to keep the exact medium, and that trusting over-much in the merciful consideration of his Maker was more acceptable to Him than the reverse, he expressed himself satisfied. He was always glad to receive from others any comforting words of this kind, looking up and attentively listening to them, and with a child-like humility regarding them for the time as his teachers, and guiding himself in his own opinions by what they said.

In allusion to his sufferings, he observed that it was a joyous thing to be under temptation : " Count it all joy when ye fall into divers temptations ;" adding, " it is not joyous at the time, but when the paroxysms are past, I rejoice that they have not drawn me from Christ." He considered sufferings as absolutely essential, and that we could not be perfected without them. " What are these pains which I endure ? As nothing compared with those of martyrs." He would remark also on the agonies of the death of the Saviour, which He endured for the whole world, and without whose interposition we should all have perished for ever : he said, " We cannot think too much about it." Yet at length he seemed almost wearied out by the

long continuance of his sufferings, and by the slow
advances of his disease. Longing to be released, he
exclaimed on one occasion, " O immortal spirit, what
a trouble it is to dispatch thee from this mortal
body ! " At another time, " How odd that I do not
die : I long for it ; " but immediately checking himself with—" Patience, patience ! "

The above remarks were mixed up with others upon
various matters, according as circumstances arose, or
as he felt himself sufficiently at ease to turn his mind
to the consideration of them. During the earlier part
of his illness he gave ample directions respecting the
disposal of his effects, and especially of his large collections of Natural History, &c. In reference to the
Ipswich Museum and the Ipswich people, he said,—
" The people of Ipswich have been very kind to me,
and have always paid me more respect than I deserve.
They will lose a president who always had the interest
of their museum at heart,—that nobody can deny. I
hope they will get another president who will take the
same interest in it, and who by living nearer will be
able to do more for them." He sometimes reflected on
his past amusements and occupations, and though quite
satisfied that they had been directed to good and useful
ends,—referring to what he had further purposed,—he
naturally compared them with the more solemn matters
which engaged his thoughts under his then circumstances :—" I thought to have a few more years
allowed me to carry out my trumpery designs," as he
called them ;—" Man purposes, God disposes."

Yet even upon his death-bed he was not insensible
to the charms of nature. It was the time of spring,

and the rooks were building in the trees in the rectory
garden, into which the window of his room looked.
"How pleasing," he observed, "to hear the rooks
cawing. God has made the world full of pleasing sights
and pleasant sounds; He might have made everything
disagreeable: it is a very good world if people will only
use it aright." His thoughts took a somewhat similar
turn on the occasion of an egg being beaten up for him
in a glass. "That egg would have become a living
being;—how full of life the world is;—what a mystery:
'In Him we live and move and have our being': the
whole world is one grand demonstration of life to
encourage us in the sure conviction of life eternal."

His love and gratitude to all about him were in-
tense, receiving with the greatest thankfulness every
little attention paid him by the family, and by the do-
mestics who waited upon him. He prayed that his
children might be a united family, showing love and
affection to each other, going through life together, re-
membering the end and the blessed mansions in which
they should all meet again hereafter. He enjoined
them to do all things in love, and to avoid selfishness.
He said many opportunities were lost—opportunities
of doing good to others as well as of advancing our
own happiness—for want of kindness and love. He,
himself, was in charity with all men. "I have no ill-
feeling (he said) to any man." He regretted on one
occasion having used a few hasty expressions with reference to some who differed with him upon religion.
He remarked, "Liberty of conscience should be allowed
to all: Christians might differ, but forbear." Similarly,
with regard to the uncharitable way in which many

persons had spoken of the authors of *Essays and Reviews,* he said he disagreed with the writers of those Essays, but did not sign the paper against them, thinking it unchristian to do so. " All men (he observed) should be allowed liberty of conscience: violent measures never do good in such a case." On another occasion he gave his opinion, that " private judgment (was) one of the greatest blessings of the land."

Talking of his past days, and the sphere of life to which he had been called, he said that amidst many dark spots, (such as all have) he was glad he had not sought after preferment. This was in allusion to a time when a friend had urged him to take steps about getting the Deanery of Ely, which was then vacant, " but he had not stirred in the matter, though he should have taken it if offered." He alluded also to the occasion before spoken of,* when there was a chance—" a toss-up," as he called it—of his being made Bishop of Norwich instead of Stanley, for which, however, he felt himself thoroughly unfit. He expressed his thankfulness that his grandfather had refused a baronetcy, as it probably would have made him a different man, and he remarked how little his father had cared for honours ; not but what they were right and proper in themselves—" Honour to whom honour is due "—only not to be coveted.

The above particulars were mostly communicated to me by his eldest son and his son-in-law, Dr. Hooker, one or other of whom was almost always with him, and who divided between them the duty of sitting up

* See *ante,* p. 142.

with him at night. Dr. Hooker was with him till
3 A.M., and observed that he always got talkative about
midnight. It was in the night time especially that
he discussed with him the subject of his past life, and
the degree to which he might take comfort from reflecting on the way in which it had been spent. His
son was in the habit of reading to him a chapter, or
the Office for the Visitation of the Sick, or some other
part of the Liturgy, in the morning when he awoke,
if he had had any sleep, or if not, whenever he felt
equal to listening to him. As the chapter was being
read he would stop and comment upon almost every
verse, sometimes for a quarter of an hour together, with
a vigour of understanding which showed how clear his
faculties still were, and how thoroughly his mind grasped
the meaning of all he heard. He would often dilate on
the different characters mentioned in Scripture, or take
occasion from the subject of the chapter to inculcate
some of the moral lessons alluded to above, or pass to
the consideration of other topics which the chapter
served to bring to his recollection. These remarks
and observations were entered by his son in a diary,
from which what has been here inserted is only a small
selection out of all that his father said at different
times for the instruction and encouragement of those
about him. Nor was this last teaching confined to his
own family. He was visited at the bedside,—before
his strength was too far gone to see them,—by many
of his parishioners, who will doubtless long remember
his death-bed advice and blessing. While to some of
his acquaintances and friends who were not within the
influence of his voice, he dictated dying words of

earnest advice and counsel, in the hope that they might feel what was uttered by him at that time. These words were taken down by Dr. Hooker at his request, and sent to the parties, among whom were some notorious evil-livers, and in one case, it is said, with good effect.

The strength and activity of his mind continued to show itself in this way at intervals for between two and three weeks from the time of his taking to his bed, after which he became so much weaker that the reading of Scripture, which he could then scarcely attend to, was gradually lessened, and at length discontinued: he spoke also but seldom, though still conscious, and able to recognize those about him.

It may be interesting, however, as well as instructive, to make a few further extracts from the diary, relating to his last days, in the order of the dates under which they are set down.

April 8th.—The Holy Communion was administered to him at 8 P.M. this day by his two sons, all the rest of the family, including four of the servants, receiving it with him. After taking the cup, he said—" O Lord Jesus, grant that these elements, accepted in simple faith and administered to me at the hands of my two sons, may be sanctified to me, may be a blessing to them—a blessing to all present; may they be preserved in the day of temptation."

April 9th.—Calm and composed, and without pain. Slept till 5 A.M., then remarked—" Nine hours since I received that blessed Sacrament. Of all the sacraments I have ever received that was the most blessed, —blessed, I trust, to all present. Saviour of this

awful world of corruption, who can tell what Thou
hast done for us sinners?"

April 10*th*.—Quiet and composed. Slept till 53.0
A.M.; a bright sunny morning, when he remarked—
"How beautiful a world this is; all love it too well
and cling to it, but I have no desire to do so now.
God has listened to one prayer, that in my last hour
I might not fall from Him. I hold fast my faith:
up to the present moment my hours have not been
clouded. I am so far allowed my hope—my trust—
my sure trust in the merits of my blessed Saviour."

April 12*th*.—On his son inquiring how he was, he
answered—"Very weak, but very comfortable, and
very happy in the hopes given me and the fears taken
from me: I have no wish to abide in this world on
my own account: if God wish it, I wish it, but not
on my own account."

April 13*th*.—Passed a good night, but spoke but
little: asked for a piece of paper on which he wrote—
"I have no dread of death, or fear of torment before
my eyes. I have perfect faith in the atonement made
by Jesus Christ for the sins of the whole world, including my own. I believe myself a partaker in His
resurrection." At 1 P.M. he had the Litany read to
him, and at 2 o'clock his old friend Professor Sedgwick
arrived from Cambridge to see him. The latter found
him "calm, resigned, and quite happy; and though
under bodily sufferings, full of peace and love." Both
were deeply affected at this last meeting. Sedgwick
stooped down and kissed his cheek; Henslow grasped
Sedgwick's hand and thanked him. Henslow then
said to Dr. Hooker, "Interpret for me, my voice is

feeble, and Sedgwick is deaf;" and so in that way
the two held conversation together for a short time.
Afterwards Sedgwick proposed instead that he should
kneel down and put his face close to Henslow's, which
he did, and Henslow put his right arm round Sedgwick's neck, and in that posture they continued talking
to each other.   When the interview was over Sedgwick
took his leave, hoping to pay him a second visit
another day, but Henslow was too weak to see him
again.   The former mentioning the circumstance of
his visit in a letter to one of the family a few days
afterwards, said that he " never saw and conversed
with a human being whose soul was nearer heaven."

April 14*th.*—Passed a good night, slept much, but
awoke apparently weaker.   He repeated his assurance
that he had no fear of death before his eyes; his feelings of assurance also that his sins were forgiven him,
and that he was going to a blissful immortality, adding
—" And if it please God that nothing should disturb
these opinions of my heart, you will all have cause to
rejoice when the hour of my dissolution is at hand."

April 15*th.*—Slept through the morning, and a
great part of the day.   In the afternoon his son found
him disturbed and hurt at reports circulated in the
village that he had come over to the revivalist opinions.
He said he wished to have such reports contradicted,
and that he had changed no religious opinions.

April 16*th.*—After passing a restless night, he
remarked, how hard it was to realize sickness without
experience : he had never experienced it before.   " Sickness (he said) should make us more gentle to those
who are sick."

April 17*th*–19*th*.—Growing weaker in mind and body: experienced increasing symptoms of oppression about the heart, which he considered as hopeful signs of death being near. He lay during these three days for thirty or forty hours together in a restless state without any sleep.

After April 21st, though he still lingered on for between three weeks and a month, he kept growing every day weaker, with occasional returns of suffering, unable to be read to, and unable to talk much, though still professing at times to be upheld by the same faith and the same love which had sustained him from the first.

On Sunday the 5th of May a further change took place, and throughout that week he spoke and ate but very little, though answering when spoken to, and recognizing those who stood by his bed: not apparently in pain, but confused as to his own state.

On Monday the 13th his son noticed that his breathing had become very irregular, several quick breaths being drawn with moaning, followed by long pauses, during which he did not appear to breathe at all. Nevertheless he was able to take a few spoonfuls of wine and water that day and the following, which was the last nourishment he swallowed. From the 14th he lay in an apparently unconscious state till the morning of Thursday the 16th of May, when, at a quarter past four, his spirit passed away so quietly that at first it was hardly known that he was actually gone.

Professor Henslow was a widower when he died, and in the 66th year of his age. He was buried on Wednesday the 22nd of May in Hitcham Churchyard, in the same vault in which his wife had been interred in the autumn of 1857. He left behind him two sons and three daughters; the sons both in orders, the eldest daughter married to Dr. Hooker, so well known in the scientific world as Assistant Director of the Royal Gardens at Kew.

He had directed the best parts of his extensive collection to be divided among the museums at Ipswich, Cambridge, and Kew. The remainder, together with his library, containing a large number of valuable works on Natural History, were sold by auction in London a few weeks after his death. His botanical diagrams—the set used in his lectures at the University—were bought by his successor in the Botanical Chair, Professor Babington.

He had " desired his funeral to be of the simplest description," and few besides his own immediate relatives met at the Rectory to accompany his remains to their last resting-place. The procession, however, had not advanced far, before it was joined by " a large number of the village labourers (allottees), who, by special request, were permitted to pay this last mark of respect to their departed pastor and friend, whom they had all learned to reverence and love. Arrived near the churchyard, the corpse was preceded by a deputation from the town of Ipswich, consisting of the Mayor and several other members of the Museum Committee, who were desirous to offer a tribute to the memory of the deceased, who had been so great a benefactor to the town and neighbourhood." Several

clergy also attended, some, it was said, who were
strangers to the place, and who had come from a distance to be present on the occasion.   The funeral
service was performed by the curate, the Rev. R.
Graves, the church and churchyard being alike filled
with the parishioners, whose mournful silence and re-
spectful behaviour testified to the impression made on
them by the solemn scene.

The grief and sympathy of Professor Henslow's
friends had been shown during his illness by the many
inquiries daily made after him from the highest to
the lowest.   It is needless to dilate on the severe loss
that was sustained by his death ; a loss felt deeply by the
parish of Hitcham, to which he had so closely devoted
himself for the period of twenty-four years, but
scarcely less felt by the University of which he was
so distinguished a Professor, and by the scientific world
in general.   Nor should mention be omitted of that
circle of attached friends who enjoyed his acquaintance, and who, while they respected his high attainments as a philosopher, yet more valued his moral
worth and the excellency of his character as a
Christian.

But when a good man dies the world does not
cease to benefit from those labours of love which he
undertook for his fellow men.   Though personally removed from them his example remains ; his voice, too,
is still heard in the lessons left to be handed down to
those who come after him.   The influences of Professor Henslow's teaching have been felt in other
places than those in which he himself taught; they

N

have already borne fruit far beyond the obscure neighbourhood in which he first sowed the good seed; and
who shall say to what further results they may not
grow in years to come, bringing honour to his memory,
and, what is far more, glory to God? " A WORD
SPOKEN IN DUE SEASON, HOW GOOD IS IT!"

LIST

PROFESSOR HENSLOW'S PUBLICATIONS,*

CHRONOLOGICALLY ARRANGED.

1821. Supplementary Observations to Dr. Berger's Account of
the Isle of Man.—(*Trans. Geol. Soc.* vol. v. p. 482.)
Geological Description of Anglesea.—(*Trans. Camb.
Phil. Soc.* vol. i. p. 359.)

1823. On the Deluge.—(*Ann. Phil.* n. s. vol. vi. p. 344.)
Syllabus of a Course of Lectures on Mineralogy. 8vo,
pp. 119.

1824. Remarks upon Dr. Berger's Reply (to Prof. Henslow's
" Observations on Dr. Berger's Account of the Isle of
Man.")—*Ann. Phil.* n. s. vol. vii. p. 407.)

1826. Remarks on the Payment of the Expenses of Out-Voters
at an University Election. Pamphlet.

1828. Syllabus of a Course of Botanical Lectures. 12mo, pp.
16.—(Later and Enlarged Editions.)
On the Crystallisation of Gold.—(*Loud. Mag. Nat. Hist.*
vol. i. p. 146.)
Botanical Museum of Cambridge, Statement respecting.
—(*Id.* vol. i. p. 82.)

1829. A Catalogue of British Plants, arranged according to
the Natural System, with the Synonyms of De Candolle, Smith, and Lindley. 12mo, pp. 40.—(2nd
Edition, 1835.)

* Including all his short communications to various scientific
journals, so far as known to the author.

1829. On the Leaves of *Malaxis paludosa.*—(*Loud. Mag. Nat. Hist.* vol. i. p. 441.)

A Sermon on the First and Second Resurrection, Preached at Great St. Mary's Church, Cambridge, Feb. 15, 1829. 8vo.

1830. On the Specific Identity of the Primrose, Oxlip, Cowslip, and Polyanthus.—(*Loud. Mag. of Nat. Hist.* vol. iii. p. 406.)

On the Specific Identity of *Anagallis arvensis* and *A. cœrulea.*—(*Loud. Mag. Nat. Hist.* vol. iii. p. 537.)

1831. On the Examination of a Hybrid Digitalis.—(*Camb. Phil. Trans.* vol. iv. p. 257.)

The Breathing-Tube of the Boa ; Snakes in the Fens of Cambridgeshire.—(*Loud. Mag. Nat. Hist.* vol. iv. p. 279.)

Rooks detecting Grubs.—(*Loud. Mag. Nat. Hist.* vol. iv. p. 280.)

The Portuguese Man-of-War.—(*Loud. Mag. Nat. Hist.* vol. iv. p. 282.)

Specific Relations of *Anagallis arvensis* and *cœrulea.*— (*Loud. Mag. Nat. Hist.* vol. iv. p. 466.)

1832. A Variety of the Common Groundsel (*Senecio vulgaris*). —(*Loud. Mag. Nat. Hist.* vol. v. p. 87.)

Fumaria Vaillantii, a British Plant.—(*Loud. Mag. Nat. Hist.* vol. v. p. 88.)

On Variations in the Cotyledons and Primordial Leaves of the Sycamore (*Acer Pseudoplatanus*).—*Loud. Mag. Nat. Hist.* vol. v. p. 346.)

On the Fructification of the Genus Chara.—(*Loud. Mag. Nat. Hist.* vol. v. p. 348.)

On the Varieties of *Paris quadrifolia,* considered with respect to the ordinary characteristics of monocotyledonous Plants.—(*Loud. Mag. Nat. Hist.* vol. v. p. 429.)

On the Specific Identity of *Anagallis arvensis* and *cœrulea.*—(*Loud. Mag. Nat. Hist.* vol. v. p. 493.)

Collectors and Collections in Nat. Hist., in the University, Town, and County of Cambridge.—(*Loud. Mag. Nat. Hist.* vol. v. p. 545.)

Review of De Candolle's "Physiologie Végétale."—(*For. Quart. Rev.* No. xxii.)

1832. On the Geographical Distribution of the Plants of Cambridgeshire.—(*Rep. Brit. Assoc.* 1832, Sects. p. 596.)

1833. Facts in Relation to the Reproductive Economy of the Mistletoe (*Viscum album*).—*Loud. Mag. Nat. Hist.* vol. vi. p. 500.)

On a Monstrosity of the Common Mignionette.—(*Camb. Phil. Trans.* vol. v. p. 95.)

1834. Some of the Habits and Anatomical Conditions of a pair of Hybrid Birds, obtained from the Union of a Male Pheasant with Hens of the Bantam Fowl; and an Incidental Notice of a Hybrid Dove.—(*Loud. Mag. Nat. Hist.* vol. vii. p. 153.)

Address to the Reformers of the Town of Cambridge.— Pamphlet.

1835. Observations concerning the Indigenousness and Distinctness of certain Species of Plants included in the British Flora.—(*Loud. Mag. Nat. Hist.* vol. viii. p. 84.)

1836. An Enumeration of Species and Varieties of Plants which have been deemed British, but whose Indigenousness to Britain is considered to be questionable.— (*Loud. Mag. Nat. Hist.*, vol. ix. p. 88.)

A Notice of the Fact, and of Particulars on the Mode, of Sugar-Candy being produced in the Flowers of *Rhododendron ponticum;* and a Notice of the Effect on the Germination of the Seeds of an Acacia by boiling them variously.—(*Loud. Mag. Nat. Hist.* vol. ix. p. 476.)

Notice of Crystals of Sugar found in *Rhododendron ponticum.*—(*Rep. Brit. Assoc.*, 1836, sects. p. 106.)

Method of Preventing the Decomposition of the Sheppey Fossils.—(*Loud. Mag. Nat. Hist.* vol. ix. p. 551.)

On the Disunion of Contiguous Layers in the Wood of Exogenous Trees.—(*Mag. of Zool. and Bot.* vol. i. p. 32.)

On the Requisites necessary for the Advance of Botany. (*Mag. of Zool. and Bot.* vol. i. p. 113.)

On the Structure of the Flowers of *Adoxa moschatellina.*—(*Mag. of Zool. and Bot.* vol. i. p. 259.)

1836. The Principles of Descriptive and Physiological Botany.
London. 12mo. pp. 322. Woodcuts.

1837. Description of two new species of *Opuntia;* with Remarks on the Structure of the Fruit of *Rhipsalis.*—
(*Mag. of Zool. and Bot.* vol. i. p. 466.)

1838. *Florula Keelingensis.* An Account of the Native Plants
of the Keeling Islands.—(*Ann. Nat. Hist.* vol. i.
p. 337.)

1840. Report on the Preservation of Animal and Vegetable
Substances.—(*Rep. Brit. Assoc.* 1840, p. 421.)
The Botanical portion of " Le Bouquet des Souvenirs."

1841. Report on the Diseases of Wheat.—(*Journ. Roy. Agric.
Soc.* vol. ii. pt. 1.)
On the Specific Identity of the Fungi producing Rust
and Mildew.—(*Journ. Roy. Agric. Soc.* vol. ii. pt. 2.)
On *Cecidomyia Tritici.*—(*Rep. Brit. Assoc.* 1841, sects.
p. 72.)

1842. Observations on the Wheat-Midge.—(*Journ. Roy. Agric.
Soc.* vol. iii. pt. 1, p. 36.)
To destroy Wasps' Nests.—(*Gard. Chron.* 1842, p. 637.)
On *Primula veris* and allied species.—(*Ann. Nat. Hist.*
vol. ix. p. 153.)

1843. On Concretions in the Red Crag at Felixstow, Suffolk.—
(*Proceed. of Geol. Soc.* 1843, p. 281.)
On Fixing Ammonia.—(*Gard. Chron.* 1843, p. 135.)
Letters to the Farmers of Suffolk, 'with a Glossary of
Terms used, and the Address delivered at the last
Anniversary Meeting of the Hadleigh Farmers' Club.
London and Hadleigh. 8vo, pp. 114.
An Account of the Roman Antiquities found at
Rougham, near Bury St. Edmund's, Sept. 15, 1843.
8vo. Pamphlet. Pp. 12.

1844. On Experimental Co-operation among Agriculturists.—
(*Gard. Chron.* 1844, p. 9.)
On the Failure of the Red Clover Crop.—(*Gard. Chron.*
1844, p. 529.)
On the Registration of Facts tending to Illustrate Questions of Scientific Interest.—(*Gard. Chron.* 1844,
p. 659.)

1844. On the action of Gypsum, with Remarks on misstatements of the writer's views.—(*Gard. Chron.* 1844, p. 691.)

Suggestions towards an Inquiry into the Present Condition of the Labouring Population of Suffolk. Pamphlet.

The Roman Tumulus, Eastlow Hill, Rougham. Opened on Thursday, the 4th of July, 1844. 8vo. Pamphlet. pp. 8.

1845. An Address to Landlords, on the Advantages to be expected from the General Establishment of a Spade Tenantry from among the Labouring Classes. Hadleigh and London. 8vo. Pamphlet.

On Nodules, apparently Coprolitic, from the Red Crag, London Clay, and Green Sand.—(*Rep. Brit. Assoc.* 1845, sects. p. 51.)

1846. Address to the Members of the University of Cambridge, on the Expediency of Improving, and on the Funds required for Remodelling and Supporting the Botanic Garden. Cambridge. 8vo. Pamphlet. Pp. 20.

Roman British Remains. On the Materials of the Sepulchral Vessels found at Warden, Bedfordshire.— (*Trans. Camb. Antiq. Soc.* 1846, 4to.)

1847. On Detritus derived from the London Clay, and deposited in the Red Crag.—(*Rep. Brit. Assoc.* 1847, sects. p. 64.)

1848. On Native Phosphate of Lime.—(*Gard. Chron.* 1848, p. 180.)

Address delivered in the Ipswich Museum, on 9th March, 1848. Ipswich. 8vo. Pamphlet. Pp. 19.

Syllabus of a Course of Lectures on Botany, suggesting Matter for a Pass-Examination at Cambridge in this subject. Cambridge. 8vo. pp. 29. (Another edition, 1853.)

On the Parasitical Habits of Scrophularineæ.—(*Ann. Nat. Hist.* 2nd ser. vol. ii. p. 294.)

1849. Parasitic Larvæ observed in the Nests of Hornets, Wasps, and Humble Bees. — (*Zoologist,* vol. vii. p. 2584.)

1849. On the Awns of Nepaul Barley (*Hordeum cœleste*, vars. *trifurcatum* and *ægiceras.—(Kew Journ. of Botany*, vol. i. p. 33.)

On the Structure of the Pistil in *Eschscholtzia Californica.—(Kew Journ. of Botany*, vol. i. p. 289.)

1850. Gnawing Power of Caterpillar of the Goat Moth.— (*Zoologist*, vol. viii. p. 2897.)

The Way in which Toads shed their Skins.—(*Gard. Chron.* 1850, pp. 373 and 422 ; also *Zoologist*, 1850, p. 281.)

Village Excursions.—(*Gard. Chron.* 1850, pp. 596, 629, 661, and 691.)

1851. Questions on the Subject-Matter of Sixteen Lectures in Botany, required for a Pass-Examination. Camb. 8vo. Pamphlet, pp. 22.

Way to Kill Fleas on Dogs.—(*Gard. Chron.* 1851, p. 749.)

Correspondence on the Clover Failure.—(*Gard. Chron.* 1851, p. 764).

On the Phosphate Nodules of Felixstow, in Suffolk.— (*Gard. Chron.* 1851, p. 764.)

On Alternate Culture.—(*Gard. Chron.* 1851, p. 779.)

1852. Food of Micro-lepidoptera.—(*Zoologist*, vol. x. p. 3358.)

Address to the Parish of Hitcham, chiefly in reference to an Attack upon the Allottees by some of the Farmers. Hadleigh. 8vo. (Pamphlet.)

1854. Note on two Brown Eagles.—(*Zoologist*, vol. xii. p. 4251.)

Suggestions for the Consideration of Collectors of British Shells.—(*Zoologist*, vol. xii. p. 4264.)

1855. On Typical Series of Objects in Natural History adapted to Local Museums.—(*Rep. Brit. Assoc.* 1855, p. 108.)

On an Anomalous Oyster-Shell.—(*Ann. Nat. Hist.* 2nd ser. pp. 314 and 385.)

1856. On Typical Forms of Minerals, Plants, and Animals for Museums.—(*Rep. Brit. Assoc.* 1856, p. 461.)

On the Triticoidal Forms of *Œgilops*, and on the Specific Identity of *Centaurea nigra* and *C. nigrescens.* —(*Rep. Brit. Assoc.* 1856. Sects. p. 87.)

Example of Botany in Village Education.—(*Gard. Chron.* 1856, p. 453.)

1856. Practical Lessons in Botany for Beginners of all Classes
Nos. 1–14.—(*Gard. Chron.* 1856, pp. 468, 484, 500,
516, 532, 565, 596, 613, 629, 676, 724, 740, 772, 837,
and 853.)

1857. A Dictionary of Botanical Terms, 12mo. pp. 218.
Woodcuts. London.

A Series of Nine "Botanical Diagrams" for the use
of Schools. Coloured Figures on Paper, 40 in. by
30 in.

The Locust in Suffolk.—(*Zoologist,* vol. xv. p. 5787.)

1858. Illustrations to be employed in Practical Lessons on
Botany ; adapted to beginners of all Classes. Prepared for the South Kensington Museum. Lond.
sm. 8vo. Pamphlet, pp. 31. Woodcuts.

1859. Celts in the Drift.—(*Athenæum,* 1859, No. 1673, p. 668.
No. 1678, p. 853.)

1860. Flint Weapons in the Drift.—(*Athenæum,* 1860, No.
1685, p. 206. No. 1721, p. 516, and No. 1723, p. 592.)

On the Supposed Germination of Mummy Wheat.—
(*Rep. Brit. Assoc.* 1860. Sects. p. 110.)

Flora of Suffolk : A Catalogue of the Plants (indigenous
or naturalized) found in a Wild State in the County
of Suffolk. By Rev. J. S. Henslow and Edmund
Skepper. Lond. and Bury St. Edmund's. Sm. 8vo.
pp. 140.

1861. Letter to the Editor of "Macmillan's Magazine," on
Darwin's " Origin of Species."—(*Macmill. Mag.* vol.
iii. p. 336.)

INDEX.

o

Woodfall and Kinder, Printers, Angel Court, Skinner Street, London.

Printed in the United States
By Bookmasters